# PHYSIOLOGUS

# PHYSIOLOGUS

*Translated by Michael J. Curley*

UNIVERSITY OF CHICAGO PRESS
CHICAGO & LONDON

The University of Chicago Press, Chicago 60637
The University of Chicago Press, Ltd., London

© 1979 Michael J. Curley
Note to the Paperback Edition © 2009 Michael J. Curley
All rights reserved. Originally published 1979
University of Chicago Press edition 2009

Printed in the United States of America

18 17 16 15 14 13 12 11 10 09      2 3 4 5 6 7

ISBN-13: 978-0-226-12870-2 (paper)
ISBN-10: 0-226-12870-9 (paper)

The woodcuts in this edition are reproduced from the 1587 G. Ponce de
Leon edition of *Physiologus*, courtesy of the Newberry Library, Chicago.

The initial research of this work was made possible in part through grants
from the National Endowment for the Humanities, a federal agency
whose mission is to award grants to support education, scholarship, media
programming, libraries, and museums in order to bring the results of
cultural activities to the general public.

Library of Congress Cataloging-in-Publication Data

Physiologus. English.
  Physiologus / translated by Michael J. Curley.
    p. cm.
  Includes bibliographical references.
  ISBN-13: 978-0-226-12870-2 (alk. paper)
  ISBN-10: 0-226-12870-9 (alk. paper)
  1. Bestiaries. I. Curley, Michael J., 1942– II. Title.
  PA4273.P8E5 2009
  883'.01—dc22

                                                    2009013135

♾ The paper used in this publication meets the minimum requirements of
the American National Standard for Information Sciences—Permanence of
Paper for Printed Library Materials, ANSI Z39.48-1992.

TO G. KARL GALINSKY

*tu duca, tu signore, e tu maestro*

CONTENTS

INTRODUCTION / *Michael J. Curley*

P *hysiologus* was one of the most popular and widely read
books of the Middle Ages. Some of the legends concerning beasts, stones, and trees which are found alle-
gorized in *Physiologus* were part of folklore as early as
Herodotus, and they have continued to exercise a strong
influence on literature and the decorative arts down to
the present day.[1] The reason is not far to seek: we all
love simple, well-told stories for their own sake. And,
if a bit of exotic lore can be mingled with useful moral
instruction, so much the better. We never stop to ask
that animal stories be zoologically accurate, for that
would be missing their point. The charm of these legends lies in their simplicity, one might even say in
their naïveté, but their popularity was assured by their
ready adaptability to a variety of cultural contexts, religious as well as secular. Beginning as Indian, Hebrew,
or Egyptian legends, they passed into Greek and Roman
folklore, poetry, and art, ultimately being absorbed into
Alexandrian handbooks of paradoxology and medical-
magical treatises.[2] From there, ancient scientific writers
such as Pliny and Aelian passed down many of these
legends to the early Christian world. The anonymous
author of *Physiologus* infused these venerable pagan
tales with the spirit of Christian moral and mystical
teaching, and thereafter they occupied a place of special
importance in the symbolism of the Christian world.
Both directly and through numerous intermediaries,
*Physiologus* became an established source of Medieval
sacred iconography and didactic poetry and was used
in the preaching manuals and religious textbooks of the
later Middle Ages.[3] The forms which the book assumed
were protean; sometimes the legends circulated independent of the allegories or the supporting Biblical citations. In one instance,[4] the allegories themselves were

copied down without the legends. The forty-odd chapters which comprised the original Greek text grew to over a hundred in some of the Latin bestiaries which were inspired by *Physiologus*. Translations of the text appeared in virtually every European vernacular, including Old English and Icelandic, providing the widest possible audience for this model *par excellence* of the allegorical method of interpreting natural history.

The following remarks are intended to introduce the reader to *Physiologus* and to place it within its intellectual and historical framework. For those who might wish to pursue the topic in greater detail, I have attached a selective bibliography.

## THE AUTHOR

The Greek word φυσιολογία (from ἡ φύσις, nature, and ὁ λόγος, word or reason) has a special meaning distinct from our English cognate "physiology." Aristotle, who was the first to use the term, speaks of certain pre-Socratic philosophers, Anaxagoras, Empedocles and Democritus in particular, as φυσιολόγοι (*Gen. an.* 4. 1. 763 31). Here and elsewhere (*De anima* 426a20), where he discusses the early theorists on the nature of sight, Aristotle uses the term to describe what we might call zoologists. But in his *Metaphysics* (986b14; 989b30; 990a3) the φυσιολόγοι are mentioned as speculative natural philosophers (again he has certain pre-Socratics in mind) who developed a theoretical system to account for the unity of the universe and the origin of being. In contrast to the Pythagoreans, who rely solely on mathematics in their philosophy, the φυσιολόγοι base their system on the observation of nature (ἡ φύσις), much as Aristotle himself did.[5] In the *Parts of Animals* (641a7), these philosophers are mentioned along with Democritus as being concerned with the formation and

causes of the shapes of animal bodies. In general, therefore, the word φυσιολόγος as first used by Aristotle was
applied to philosophers whose theories about natural
phenomena were grounded in observation of the sensible world and, in particular, of animal life.

A rather different meaning is attached to the term in
Diodorus Siculus and in Cicero. Describing the honors
accorded the Tyrrhenians (i.e., the Etruscans), Diodorus
notes that these people excelled in the perfecting of letters (γράμματα), φυσιολογία, and theology. He then
proceeds to explain that they elaborated the craft of
divination, being especially gifted in interpreting the
omens of Zeus through thunder and lightning.[6] This
association of the term with knowledge of the secret,
occult operations of the natural world explains why it
can be coupled with theology. I assume that to Diodorus
γράμματα would include incantations and religious-
magical formulas, for in no other kind of "letters" were
the Etruscans known to excel. Cicero, Diodorus's Latin
contemporary, shows that the meaning of φυσιολογία
had acquired a new range of meaning since Aristotle's
day. In his *De divinatione* (1. 40. 90) Cicero's brother
Quintus, speaking about the practice of divination
among the Gauls, mentions that he has firsthand knowledge of the topic, through personal acquaintance with a
Druid named Divitiacus. "He claimed to have that
knowledge of nature which the Greeks call 'physiologia,' and he used to make predictions, sometimes by
means of augury and sometimes by means of conjec-
ture." By the beginning of the Christian era, in other
words, in both the Latin and the Greek worlds, φυσιολογία had acquired a range of meaning extending from
early Greek speculative zoology and physics to the occult religious practices of exotic peoples.

Among the ancients, however, it is to Plutarch's treatise *On Isis and Osiris* that we must turn to find the first

explicit association of animal symbolism with theologi-
cal speculation. Plutarch claims that the Egyptians based
their sacred symbolism on natural phenomena, which
supposedly contained esoteric teachings about the nature of the gods. He mentions, for example, that the
crocodile is honored among the Egyptians as the living
representation of God since he is the only creature
without a tongue, "for the divine word has no need of a
voice" (φωνῆς γὰρ ὁ θεῖος λόγος ἀπροσδεής ἐστι),[7] and
that he is endowed with a nictating membrane allow-
ing him, like God, to see without being seen (ὥστε
βλέπειν μὴ βλεπόμενον, ὃ τῷ πρώτῳ θεῷ συμβέβηκεν).[8]
In a later passage, Plutarch compares the Greek custom
of representing divine qualities in statues and in num-
bers and figures (as among the Pythagoreans) with
Egyptian animal symbolism:

If, then, the most noted of the philosophers, observing
the enigma of the divine (αἴνιγμα τοῦ θείου) in inani-
mate and incorporeal objects, have not thought it proper
to treat anything with carelessness or disrespect, even
more do I think that, in all likelihood, we should wel-
come those peculiar properties existent in natures which
possess the power of perception and have a soul and
feeling and character. It is not that we should honor
these but that through these we should honor the Divine, since they are the clearer mirrors of the Divine
by their nature also, so that we should regard them as
the instrument or the device of the God who orders all
things. . . . The nature that lives and sees and has within itself the source of movement and a knowledge of
what belongs to it and what belongs to others has drawn
to itself an efflux and portion of beauty from the Intelligence "by which the Universe is guided" (ἀπορροὴν
καὶ μοῖραν ἐκ τοῦ φρονοῦντος, "ὅτῳ κυβερνᾶται τὸ
σύμπαν"), as Heracleitus has it. Wherefore the Divine is
no worse represented in these animals than in works of
bronze and stone, which are alike subject to destruction
and disfiguration and by their nature are void of all per-

ception and comprehension. This, then, is what I most approve in the accounts that are given regarding the animals held in honor.[9]

Somewhat later, under the influence of Neoplatonic currents during the first two centuries A.D., Christian writers conceived of a more systematic φυσιολογία with a mystical dimension nowhere to be found among the ancients. Clement of Alexandria, for example, in book 4 of his *Stromata* speaks of a γνωστικὴ φυσιολογία, by which he appears to mean an initiation into the knowledge of the heavenly mysteries by way of their earthly correspondences (τὰ μικρὰ πρὸ τῶν μεγάλων μυηθέντες μυστηρίων).[10] Perhaps the clearest statement of the principle on which such a mystical φυσιολογία is based, however, can be found in book 3 of Origen's *Commentary on the Song of Songs*, the entire second half of which is dedicated to exploring the analogical structure of nature. Origen writes,

The apostle Paul teaches us that the invisible things of God may be known through the visible (*invisibilia Dei ex visibilibus intelligantur*), and things which are not seen may be contemplated by reason of and likeness to those things which are seen. He shows by this that this visible world may teach about the invisible and that earth may contain certain patterns of things heavenly, so that we may rise from lower to higher things (*ut ab his, quae deorsum sunt, ad ea, quae sursum sunt, possimus adscendere*) and out of those we see on earth perceive and know those which are in the heavens. As a certain likeness of these, the Creator has given a likeness of creatures which are on earth, by which the differences more easily might be gathered and perceived. And perhaps just as God made man in his own image and likeness, so also did he make the remaining creatures after certain other heavenly images as a likeness. And perhaps every single thing on earth has something of an image and likeness (*habent aliquid imaginis et similitudinis in caelestibus*)

in heavenly things, to such a degree that even the grain
of mustard which is the smallest of all seeds may have
something of an image and likeness in heaven.[11]

When the author of *Physiologus* says that we should
understand the "intelligible" pearl (*intelligibilis marga-
rita*) or elephant or dove, he is relying on Origen's concept of the correspondences between the phenomenal
world of nature and its heavenly archetype, of which it
is a likeness (*similitudo*). In other words, Clement and
Origen allowed the allegorical method of Philo (with the
authority of Saint Paul) to be applied to the "book of
nature."[12] As Paul said, "Ever since the creation of the
world his invisible nature, namely, his eternal power
and deity, has been clearly perceived in the things that
have been made" (Romans 1:20). Under the conviction
that the Creator had infused nature with his traces, the
Alexandrian allegorists devised a mystical hermeneutics
by which man's limited terrestrial understanding could
embrace the divine. Although their theologies were very
different, the Christian and pagan allegorists, especially
in Egypt, shared much common ground in discerning
within the operations of the natural world "a portion of
the beauty of the Intelligence by which the universe is
guided." Both systems exploit certain natural processes
as a symbolic language whose vocabulary itself bears an
explicit correspondence to the nature of the Creator. For
a Christian Platonist such as Origen, however, these
*imagines* and *similitudines* were deliberately fashioned by
God to lead man's weakened intelligence to a knowledge of things hidden and spiritual. While we would
not be justified to speak here of an early Christian
"semiotics" of nature, since nothing so elaborate was
envisioned by Clement or Origen (or by Plutarch for
that matter), it is true to say that the idea of the ana-
logic structure of nature furnished the Christian world
with a ready set of divinely provided symbols to bridge
the otherwise impassable intellectual gulf between the

*visibilia* of this world and the *invisibilia* of the other.

The aim of this method was far different, too, from that of a Sophist naturalist such as Aelian (fl. A.D. 200), for whom the behavior of animals is often a model of human conduct, an opportunity for a sermon on the social harmony of the natural world, or a lament on man's inhumanity to man; his purpose is entirely ethical and social. *Physiologus*, to be sure, also elicits purely moral lessons from the natural world but more often aims at making manifest the nature of God himself by unveiling the vestiges of the Creator in creation. This is the justification for a Christian φυσιολογία.

Finally, in addition to the moral-mystical significance which early Christian authors gave to the ancient idea of φυσιολογία, the term was also used by writers of the early Christian period to mean simply any allegorical interpretation of a myth or legend. Thus, Tertullian in his tract *Ad Nationes* interprets *physiologice per allegoricam argumentationem* Saturn's devouring his children to mean that Time consumes everything that arises from it.[13] Somewhat later, Servius, in his commentary on line 295 of book 6 of Vergil's *Aeneid*, speaks of his moralized allegory of the geography of Hades as a *physiologia*.[14] The mythographer Fulgentius claimed to have written a book entitled *Physiologus*, but its precise contents are unknown to us.[15]

I have lingered over the meaning of *physiologia* among the ancients and early Christians to dispel the notion that the title of our book means "the naturalist," as we often read. *Physiologus* was never intended to be a treatise on natural history. For that purpose, the Middle Ages had Pliny. Nor did the word φυσιολόγος ever mean simply "the naturalist" as we understand the term, as should now be evident, but one who interpreted metaphysically, morally, and, finally, mystically the transcendent significance of the natural world.

Physiologus, no doubt, was originally understood to

be a person, not the title of a book. The formulas *Physiologus dicit* and *Bene Physiologus dixit* show that a person
is meant. Who was he? We will probably never know.
The one thing that seems certain from what has been
said, however, is that, to have merited his name, he
must have written about the natural world in the spirit
of what we know φυσιολογία to have meant. In any
case, throughout the Medieval period Aristotle and
Solomon, the Greek and Hebrew interpreters of nature
*par excellence*, were alternately identified as Physiologus.
The Byzantine recension of *Physiologus* makes Solomon
the author of the legends and Saint Basil the author of
the allegory.[16] Among the host of other noteworthy authorities proposed at one time or another as the author
of *Physiologus*, we might mention Peter of Alexandria,
Epiphanios, John Chrysostom, Athanasius, Ambrose,
and Jerome.

## PLACE OF ORIGIN

I t has generally been accepted that *Physiologus* was originally produced in Egypt and, more particularly, in
Alexandria for the following reasons:

1. A number of legends in *Physiologus* can be traced
back to Egyptian folk beliefs and religious worship concerning sacred animals. The phoenix was, of course, the
sacred bird of the city of Heliopolis, where it was wor-
shipped as part of the cult of the sun god. The ichneumon, the niluus (or hydrus), and the crocodile were
associated among the Egyptians with legends concerning the Nile and the battles of the forces of light and
darkness.[17] The chapter on the wild ass and the monkey
appears to rely on an Egyptian folk belief about the in-
auspiciousness of the twenty-fifth day of the month of
Phamenoth. This and another Coptic month, Pharmouthi, are also mentioned in the chapter on the
phoenix.

2. Some fifteen legends are common to *Physiologus* and the treatise on hieroglyphics known as the *Book of Horapollo*. The similarity of parts of our text to the *Kyranides* of "Hermes Trismegistos" can be explained by the direct influence of the former on the latter and, partly, by a common dependence on a reservoir of popular Egyptian lore.

3. The allegorical method of *Physiologus* ultimately reaches back to the Hellenized Jewish scholars of the schools of Alexandria and their most eloquent spokesman, Philo, from whom Christian exegetes, particularly Clement of Alexandria and Origen, derived their manner of interpreting Scriptures. The indebtedness of our text to Alexandrian hermeneutics has never been in doubt. Scholars (such as Max Wellmann) who think that *Physiologus* was composed in Syria by a disciple of Origen[18] acknowledge its Alexandrian flavor.

4. The existence of an early Ethiopian translation of *Physiologus*, according to its editor and translator Fritz Hommel, points to Alexandria as a source, since the overwhelming majority of the earliest Ethiopian texts are derived from Alexandrian models.[19]

5. The well-established tradition of Alexandrian paradoxography, i.e., popular literature concerning marvels ($\pi\alpha\rho\acute{\alpha}\delta o\xi\alpha$, $\theta\alpha\upsilon\mu\acute{\alpha}\sigma\iota\alpha$), dates back at least to Callimachus and his contemporary, Bolos. While *Physiologus* is not primarily a work of paradoxography, it nevertheless shares common ground with that form of literature in its love of exotic subject matter and fabulous natural history.

THE DATE

The date of the original Greek version of *Physiologus* is much debated. Lauchert, who wrote the most complete study of the work, was of the opinion that it was in circulation around A.D. 140 or even during the first quarter of the second century.[20] The evidence for such an early date consists of what Lauchert thought to be echoes of *Physiologus* in the writings of the Greek church fathers. One of these is especially important. In Origen's seventeenth homily on Genesis 4:9, the benediction of Jacob ("Judah is a lion's whelp; from the prey, my son, you have gone up. He stooped down, he crouched as a lion, and as a lioness; who dares rouse him up?") the following passage appears: "A mystical exposition is provided for this passage in which the lion's whelp represents Christ not only figuratively (*tropice*) but truly also by nature (*physice*), for Physiologus wrote the following about the lion's whelp: after being born, it sleeps for three days, and three nights after which the sire's growling and roaring makes the lair tremble and awakens the sleeping whelp. That other whelp, therefore, arose from the shoot, for He was born of a Virgin." Now if Origen (ca. 185–254) did write this passage, it would be our earliest explicit mention of *Physiologus*. But it is now generally accepted that Rufinus, Origen's Latin translator, interpolated this and many other such passages into the original homily sometime between A.D. 397 and 410. The echoes that Lauchert perceived in the works of Clement of Alexandria (ca. 150–215), Justin Martyr (ca. 100–165), and the pseudo-Clement of Rome (*The Recognitions*, 211–231) cannot be received as anything more than faint reverberations of a general tradition of folklore from which they and our author commonly drew.

At the other end of the spectrum are those scholars who suggest that *Physiologus* was originally composed sometime during the fourth century. Karl Ahrens, for example, thought that the warnings against heresy which are found in the Greek version of the legend of the ant, the sirens, the whale, and the coot are explainable in the light of fourth-century church history, especially the turmoil created by the Arian heresy.[21] Such an

argument, however, is not entirely convincing, since
many early Christian writers such as Justin Martyr often
refer to heretics. Besides, it would not be correct to claim
that *Physiologus* displays notable concern over the prob-
lem of heresy. In fact, the faint traces of Gnosticism
discernible in the chapters on the lion, the unicorn, and
the adamant-stone argue for an early date of com-
position, when such incipiently "heretical" doctrines
were to be found among the most orthodox writers,
not to mention the Scriptures themselves. It was not
until 534 at the Synod of Constantinople that the Gnostic doctrine of the Incarnation, of which there are traces
in *Physiologus*, was finally condemned as one of the nine
errors of Origenism.[22]

On the other hand, Wellmann has clearly demon-
strated that toward the end of the fourth century *Physiologus* was quoted verbatim for the first time by the
author of the Hexaemeron commentary of Eustathios.[23]
Also about this time in the Latin tradition (ca. 386–388),
Saint Ambrose seems to draw directly on the wording
of *Physiologus* in his *Hexaemeron* when telling the legend
of the partridge. This passage in Ambrose is especially
important since the partridge story does not appear in
literature earlier than *Physiologus*. That Ambrose recounts both the legend and the allegory as they are
found in our text makes it quite likely that he knew
*Physiologus* at first hand.[24]

The earliest translation of the Greek text appears in
Ethiopian toward the beginning or middle of the fifth
century A.D., and not long afterward the Syrian and
Armenian versions were produced.[25] On purely textual
grounds F. Sbordone concluded after studying seventy-
seven Greek manuscripts and their genealogies that the
original Greek text was composed around the year A.D.
200.[26] We can conclude, therefore, that the Greek text
was in circulation by the last quarter of the fourth century; it may have been composed as many as two hun-

dred years earlier, although evidence for this opinion is by no means conclusive.

The earliest Latin translation of *Physiologus* could have appeared sometime in the fourth century, if Ambrose (who knew Greek) was quoting from a Latin text in the passage on the partridge in his *Hexaemeron*. The allegory on the third nature of the ant in one family of Latin manuscripts warns against certain heresies: "Flee, O man of God, barley, that is the teachings of the heretics, for they are barley and harmful things fit to be cast among rocks. Heresies kill the souls of men. Flee Sabellius, Marcion, Manichaeus, avoid Novatianus, Montanus, Valentinus, Basilides, Macedonius, avoid Donatianus and Photinus and all who come forth from the Arian brood like serpentine offspring from the womb of the dragon. The dogmas of these men are false and hostile to truth." Noticing that the name of Nestorius is absent from this list, Lauchert proposed that the Latin text must, therefore, have been composed ca. 397–431, since Nestorianism was condemned at the Council of Ephesus in 431.[27] While such a date for the Latin translation is not inherently improbable, the argument from omission is always open to some doubt. Perry, in fact, has pointed out that in some Latin versions of *Physiologus* the names of Montanus and Macedonius are also lacking and that this can be explained by scribal inattention or tampering as well as by chronology.[28] Sbordone's study of the genealogy of the manuscripts led him to conclude that the Latin translation goes back to one of the oldest Greek texts, even though the earliest surviving Latin manuscripts date from the eighth (MS Bern 233), ninth (MS Bern 318), and tenth centuries (MS Wolfenbüttel Cod. Gud. 148 and MS Bruxells 10,074). The first explicit reference to a Latin *Physiologus* occurs in the fifth part of the so-called *Decretum Gelasianum*, entitled "De libris recipiendis et non recipiendis." Among the books to be shunned is one "*Liber Physi-*

*ologus* ab hereticis conscriptus et beati Ambrosii nomine praesignatus, apocryphus."[29] The *Decretum Gelasianum* purports to be an official document of Pope Gelasius (492–496); it is, however, strongly suspected of being a private compilation dating from early sixth-century Italy.[30] That someone during this period should think *Physiologus* a heretical work is perhaps not as important as the fact that subsequent orthodox writers openly ignored the prohibition (if they ever knew of it) against reading it. The Latin translation, therefore, was known in the early sixth century and quite possibly was circulating as early as the mid fourth century.

## THE SOURCES

In keeping with the general uncertainty that clouds the historical background of *Physiologus*, we know of no single "source" which provided our author with the material for his work. This fact is not altogether surprising, since a glance at the analogues provided in the notes following the translation will show that *Physiologus* is a syncretic work which draws upon folk legends and pseudoscience common to many Eastern Mediterranean cultures—Greek, Roman, Egyptian, Hebrew, and Indian. Some of these tales certainly date back to prehistory while others, as far as we know, appear in their classic form for the first time in the pages of *Physiologus* itself.[31] Pliny's *Natural History* records many stories closely analogous to those of our text, and Aelian's *History of the Animals*, as has already been noted, attaches similar moral interpretations to animal behavior, real or imagined. Max Wellmann has shown that the magical-medical properties of animals, stones, and trees as described in *Physiologus* can be traced back to the writings of Bolos of Mendes (third century B.C.), as can the sympathy-antipathy framework of many of the legends in *Physiologus*.[32]

Some early students of *Physiologus* regarded the book as consisting of two distinct parts: the legends and the allegories. The existence of a number of versions containing only the legends led some scholars to conclude that the two parts were in fact written independently.[33] The legends, according to this view, were first compiled into a pseudoscientific medley by a pagan author who drew on the well-known literary sources of ancient natural history from Herodotus and Aristotle down through Pliny and Aelian. To this collection, a Christian writer familiar with the allegorical exegesis of the Alexandrian school attached the interpretations as we now have them. The intention of the author was to render *physiologice per allegoricam argumentationem* the true kernel of meaning within the integument of a pagan fable. Or, to paraphrase Saint Paul, the author set about to let us see face to face what the ancients perceived only through a glass darkly. As we have seen already, this view had its advocates during the later Middle Ages among those who thought that Solomon or Aristotle was responsible for the legends and that Saint Basil was the author of the allegories. The general view was that the author played a passive role in collecting the preexistent legends and an active one in composing their interpretations. There is something to be recommended in this view. The legend of the beaver, for example, is told among other places in Pliny's *Natural History* (8. 109): "The beavers of the Black Sea region practice self-amputation of the same organ [i.e., their genitals] when beset by danger, as they know that they are hunted for the sake of its secretion, the medical name for which is beaver-oil." Now the ancient poets were not slow to perceive a moral in that story. Juvenal uses the beaver as a symbol of prudence in *Satire* 12. 29f., where he portrays the wealthy Catullus in the midst of a potential shipwreck jettisoning his luxurious possessions into the sea to preserve his life at the expense of his cargo, just as the wise beaver casts off his genitals to preserve his life. This story survived scientific protestations against its accuracy and was repeated by Herodotus and Aesop among the Greeks and by Cicero and Apuleius among the Latins.[34] Now the version of the beaver's self-castration which appears in *Physiologus* does not appear to have been altered substantially from its ancient analogues. The animal's mildness (*innocentissimum valde et quietum*) is mentioned by way of assimilation with the previous chapter on the unicorn. The mention of his second escape from the hunters has been borrowed from Aelian's version (*On Animals* 6. 34), where the detail first appears. For the most part, however, our author found the traditional story consonant with Christian morality and hence in no need of manipulation before "attaching" the allegory. The beaver is every true Christian, the hunter is the devil (a common metaphor in the Middle Ages), the beaver's genitals are the inclinations toward sin in all of us, and the entire legend illustrates Saint Paul's admonition in Romans 13:7, "Pay all of them their dues, taxes to whom taxes are due, honor to whom honor is due, etc." With a deft application of hermeneutics, an ancient legend is transformed into a Christian fable. Since the details of the ancient legend are left largely intact in *Physiologus*, we can properly speak of the allegory in this case being appended to a pagan model. In most cases, however, the author plays a far less passive role in transmitting the conventional legend.

A good example of how wholly new details insinuate themselves into *Physiologus*'s legends can be found in the description of the first nature of the serpent. When this animal grows old, its eyes become dim. Wanting to renew itself, it fasts for forty days and forty nights; when its skin has become loosened from its flesh, it finds a narrow crack (*fissura angusta*) in a rock and, squeezing itself through the crack, casts off its skin and

is thus rejuvenated. Now a very similar legend is told
of the serpent in the ancient analogues, especially in
Pliny and Aelian.[35] There are some revealing differences, however: nowhere does either ancient writer
mention that the serpent "fasts" for forty days and
nights. Pliny states in one passage (8. 59) that a serpent
can endure a hibernation of a year without food, provided that it is protected against the cold. Elsewhere
(8. 41) he says that the snake's sloughing requires at
least twenty-four hours to complete. The fast of forty
days, however, was suggested to our author by Matthew 4:2, where Christ is said to have fasted forty days
and forty nights in the desert immediately following his
baptism and just before his temptation by the devil.
Similarly, although both Aelian and Pliny mention the
serpent's hibernation in a hole, neither suggests that the
sloughing occurs as a result of the animal's emergence
through a "narrow crack" in the ground. This embellishment is entirely the work of a Christian author purposely echoing the language of Matthew 7:14, "The gate
is narrow ($\H{o}\tau\iota$ $\sigma\tau\epsilon\nu\grave{\eta}$ $\H{\eta}$ $\pi\acute{\upsilon}\lambda\eta$) and there is tribulation on
the way which leads toward life, and few are those who
enter through it." On one level, the legend of the
snake's rejuvenation has been metamorphosed into a
quadragesimal allegory of one of the episodes in the life
of Christ. On another level, it represents the struggle of
the soul of every Christian to overcome the burden of
the flesh and to attain a state of purification. Again,
only in *Physiologus* do we find mention of the old age of
the snake. Sbordone[36] suggests that this detail arises
from a misunderstanding of the Greek expression
$\dot{\alpha}\pi o\beta\alpha\lambda\epsilon\hat{\iota}\nu$ $\tau\grave{o}$ $\gamma\hat{\eta}\rho\alpha s$, meaning "to slough off a vestigial
layer of skin" ($\tau\grave{o}$ $\gamma\hat{\eta}\rho\alpha s$), and not "to put off or cast off
old age" (since $\tau\grave{o}$ $\gamma\hat{\eta}\rho\alpha s$ can also mean "old age"). From
what has been said thus far, however, it seems more
likely that this departure from the analogues is a delib-
erate intervention on the part of the author to create an

"enigma of the Divine" where nature and art provided none. And what of the serpent's blindness? It is cured according to the ancients by the beast's rubbing its eyes against the fennel plant, of which there is no mention in *Physiologus*. The absence of this therapeutic plant introduces some doubt into the reader's mind whether the beast remains blind following its sloughing or its eyesight is restored with its new skin. Nevertheless, there can be little doubt that the language of the entire first nature of the serpent is reminiscent of Saint Paul's words to the Colossians 3:9 and the Ephesians 4:22: "Put off your old nature . . . and put on the new nature created after the likeness of God in true righteousness and holiness." In working to enhance the legend's moral and mystical content, our author sacrifices some of the scientific details of the tradition, leaving a narrative rough edge here and there. But his general purpose remains clear: to conform the "facts" of the visible operations of nature to their "intelligible" archetypes in accord with Christian φυσιολογία.

Not only are the details of many legends purposely altered to harmonize them with Christian doctrine, but the very substance of some stories has been subjected in *Physiologus* to a radical metamorphosis for the same purpose. Ancient authorities tell us that the panther (in Aelian it is the leopard, ἡ πάρδαλις) gives off a marvelous fragrance which attracts all animals.[37] Because he is a fearsome beast to behold, he lurks in the underbrush unseen by his unsuspecting prey as they are irresistibly drawn to his sweet odor. Once within range, he springs upon them suddenly and devours them. Although the details vary from author to author, all the ancients agree that the panther is a predator who uses his fragrance as a deadly enticement.[38] He is, in other words, an example of one of the "antipathies" of the natural world, since he is an enemy of all other beasts. The story as we find it in *Physiologus*, however, has undergone a num-

ber of interpolations calculated to turn the animal's traditional symbolic value from one of universal antipathy
to one of universal sympathy (excepting the hostility
toward the dragon, of course), thus reversing the original point of the legend. The panther now becomes a
mild animal (*quietum . . . et mitissimum*) whose coat is
beautiful and variegated. Awakening from sleep after
three days, he roars and breathes forth a pleasant fragrance (*odor bonus aromatum*) which causes all beasts far
and near to follow him. In this case, the legend rever-
berates with echoes of the New Testament. No phrase is
without allegorical significance, even such parenthetical
phrases as "those who are far and near" (*qui longe sunt
et qui prope*), by which we are to understand the Gentiles
who are far away without the Law and the Jews who
are near living within the Law.

Many other chapters could be cited in detail to show
how the author of *Physiologus* has reshaped the traditional material of his analogues to conform to a pre-
conceived allegory. Sometimes this refashioning is
accomplished at the expense of narrative logic, as we
have seen in the case of the blindness of the serpent.
But, in most cases, the author has succeeded in retelling
the old tales with the spirit *and* letter of the Christian
Scriptures. Behind *Physiologus* there may have been a
pseudoscientific anthology from which our author derived his lore, but there can be little doubt that in transmitting these legends he transformed them in the interest of Christian doctrine and within the conceptual
framework of a Christian Neoplatonic φυσιολογία.

## THE LATER HISTORY OF *Physiologus*

The influence of *Physiologus* on the literature and art of
the later Middle Ages is too long a story to be fully recounted here. Its impact on literature begins with the
earliest translation of the Greek text into Ethiopian to-

ward the beginning or middle of the fifth century. The value of the Ethiopian translation, apart from its being one of the earliest literary works extant in that language, is its apparent fidelity to the Greek original.[39] Probably as old as the Ethiopian version is the Armenian, which stems from a fourth-century Greek manuscript; it agrees closely with the Greek in most of the legendary details, although the allegories show marked variation from the original. The Armenian translation contains thirty-five beasts, of which three appear in no known Greek model (the Zerehav bird, the bee, and the tiger). As old as the Ethiopian and Armenian versions is the Old Syrian translation, in which the allegories and in some cases the Biblical citations are abandoned. Although the Syrian translator appears to have understood Greek better than his Ethiopian and Armenian contemporaries, he did not hesitate to draw upon sources outside *Physiologus* in at least one instance, the story of the ostrich. The impulse to expand the inventory of *Physiologus* begins in these early translations. Land's edition of the Syrian version, for example, contains eighty-one chapters, of which only the first forty-seven are taken from *Physiologus*; the rest have been borrowed from Basil's *Hexaemeron*. It is also at this time that attempts to identify the author begin. The title of Land's edition reads, ". . . liber Physiologi sive verba facientis de naturis, et compositus est a S. Basilio, cum theoria sive expositione ex illis deprompta." This tradition of attribution was followed in later centuries when authorities such as Ambrose, Chrysostom, Epiphanios, and Jerome, all of whom wrote on natural history and were often indebted to *Physiologus*, were proposed as authors of our book. There are two Arabic translations of *Physiologus*, one containing only eleven chapters, the other only thirty-five. The latter translation follows the Ethiopian order of chapters and probably reaches back to a very early Greek model. Its heading attributes the original to

"Gregorius Theologus" (i.e., Gregory of Nazianzus). Probably during the eighth century a Georgian translation appeared which was the archetype (now lost) of the surviving tenth-century Georgian version.

The earliest Latin *Physiologus*, as indicated above, appears sometime between the fourth and the sixth centuries, although the earliest surviving manuscripts date from the eighth, ninth, and tenth centuries. The Latin translations are usually characterized by an expanded allegory (see chapter XXIV, "On the Oyster-stone and the Pearl"). Association of the elephant with the Samaritan, the interpretation of the many colors of the doves, and the independent chapter on the viper all arise from the enlarged allegories in the Latin texts.

Excerpted versions of *Physiologus* begin to appear with the so-called alphabetical *Glossary of Ansileubus*, which omits the allegories.[40] Similarly, the so-called *Dicta Chrysostomi* (see Lauchert, pp. 92–94), from which many of the German translations stem, contains only thirty-two chapters and entirely omits stones and plants from its inventory. The earliest manuscript of the *Dicta Chrysostomi* dates from the eleventh century, although Chrysostom was known as the author of a "liber de natura" as early as the sixth century. From the *Dicta Chrysostomi* arises the so-called *Theobaldus-Physiologus*,[41] a metrical version probably of the eleventh century which describes thirteen animals, including an original chapter on the spider. This version was very popular during the Middle Ages, when it was used as a schoolbook complete with glosses and commentary. No doubt, when certain authors during the late Middle Ages quote *Physiologus* as an authority, they are thinking of the author they read as schoolboys, who in many cases was Theobaldus.

The earliest European vernacular version of *Physiologus*, and surely the most inspired, is the Old English poetic text dating from the second half of the eighth century and sometimes attributed to Cynewulf.[42] Three beasts are described—the panther, the whale (called the "Fastitocalon"), and a bird, probably the partridge—for which only sixteen lines of the allegory are preserved. Perhaps the Old English poem as we have it is only a fragment of a complete translation of *Physiologus*, since the three beasts described appear in exactly this order in the Latin text. Quite possibly, however, this selection represents a conscious attempt on the part of the Old English poet to create a triptych of nature displaying the powers of the denizens of land, sea, and air. In any case, the poet's exquisite embroidery on the traditional legends demonstrates his independence from the received material. This same freedom of interpretation with regard to *Physiologus* is characteristic of its reception during the Middle Ages. Just as Physiologus itself, as we have seen, was a product of diverse tributaries, so too its branches in the course of its long journey through history never preserve the exact contours of the source. This was a book of inspiration as well as information. In Old English of about 750–800 is another poem attributed to Cynewulf on the subject of the phoenix, but its source is Lactantius's poem *De ave phoenice.* Finally, two Old High German translations of the *Dicta Chrysostomi* version date from the eleventh and twelfth centuries, and we also have fragments in Flemish, Icelandic (of the thirteenth century), Provençal, Russian, and Medieval Greek.

A major watershed in the later history of *Physiologus* is its incorporation into the encyclopedias and natural-historical compendia of the late Middle Ages. Isidore of Seville's great *Etymologies* (ca. 623) transmitted an enormous volume of animal lore in book 12, entitled "De animalibus." Isidore's sources included an impressive number of ancient authors and church fathers, from whom he derived much material common to *Physiologus.* He omits the allegories, however. In addition to

Isidore, the late Middle Ages came to know the animal legends originally popularized by *Physiologus* in the influential works of Thomas of Cantimpré, Albertus Magnus, Vincent of Beauvais, the pseudo–Hugh of Saint Victor, Alexander Neckam, Jacques de Vitry, Bartholomeus Anglicus, and Honorius of Autun.

By the end of the twelfth century a new form of popular nature-book had developed under the generic name of "the bestiary" which, in keeping with the encyclopedic taste of the period, tended to absorb virtually all animal legends, including those of *Physiologus*, into its pages. While the Latin text of *Physiologus* contains about fifty chapters, the greatly expanded bestiaries contain as many as one hundred and fifty chapters. The bestiaries are frequently divided into zoological categories of mammals, birds, fish, reptiles, etc., and often carry over the etymological learning of Isidore, informing us, for example, that the viper (*vipera*) is so named because it gives birth with violence (*vi*), that the phoenix derives its name from its purple color (*phoeniceus*), and that the beaver is called *castor* because of its self-castration. In many instances, chapters derived from *Physiologus* in the bestiaries still begin and end with the formulas, "Physiologus says of the . . ." and "Physiologus spoke well of the . . ."

It is true to say that direct knowledge of *Physiologus* was somewhat diluted by the enormous popularity of Isidore and the bestiaries among writers and of Honorius among artists. Most people, however, credited to *Physiologus* whatever animal lore was not explicitly attributable to Pliny, Aesop, Phaedrus, and Avienus. In most cases, in fact, it is very difficult to prove a single "source" for a given representation of an animal legend during the late Middle Ages. Yet we should keep in mind that, whatever the intermediaries, *Physiologus* was responsible ultimately for some of the most enduring iconography of Christianity. The symbolic lion, unicorn, panther, phoenix, whale, and pelican, which are so familiar to us, all trace their ancestry back to our little book.

In addition to Latin works, *Physiologus* had a forceful impact on the popular French bestiaries produced between 1125 and 1225 by Philippe de Thaon, Gervaise, Guillaume le Clerc, and Pierre de Beauvais. According to Florence McCulloch, most French bestiaries can be shown to derive from a version of *Physiologus* containing much material borrowed from Isidore. The oldest French bestiary (3,194 lines) was composed by the Anglo-Norman poet Philippe de Thaon ca. 1121 and dedicated to Aelis de Louvain, the second wife of Henry I. This work contains thirty-eight chapters treating beasts, birds, and stones in that order. Philippe cites as his sources *Phisiologus bestiaire* or *bestiaire, un livre de gramaire, Ysidre* (i.e., Isidore), and *Escripture*, though it is probably his immediate Latin model rather than direct knowledge that accounts for these sources. The importance of Philippe's translation lies not so much in its inherent literary qualities but, as Florence McCulloch states, "in the fact that a tradition already ancient and rich had now entered the vernacular to become widely read and known in the next century and a half."[43] Gervaise's rhymed *Bestiaire* (1,280 lines) was probably composed toward the beginning of the thirteenth century and follows the *Dicta Chrysostomi* version of *Physiologus*. Perhaps the most literary of the French bestiaries is that of the early thirteenth-century Norman poet Guillaume le Clerc or Guillaume le Normand, whose work consists of some 3,426 octosyllabic lines. Guillaume's poem was copied repeatedly and illustrated amply down through the fifteenth century and must have been one of the principal vehicles in the vernacular for transmitting the lore of *Physiologus* to the later Middle Ages. Pierre de Beauvais's bestiary, composed before 1218, exists in two renditions—a short version of twenty-eight chapters and a long version of about seventy-one chapters, the latter in the Picard dialect.

Finally, mention should be made of the fourteenth-century Waldensian bestiary containing fifty-four chapters, of which some twenty-five derive from *Physiologus*. [44] This collection appears to have been composed for student use by one Jaco and is a very free translation, drawing on a number of sources outside *Physiologus*. In keeping with a common late Medieval attitude to the allegories, the Waldensian version tends to replace the mystical interpretations of the original with moral ones. Italian bestiaries appear in Tuscan during the mid thirteenth century, while the *Bestiario moralizato*, a sonnet sequence, appears slightly later. The *Fiori di virtú* dates perhaps from the beginning of the fourteenth century and contains a dozen chapters, out of a total of thirty-five, from *Physiologus*. Having been translated into Greek, Armenian, Rumanian, Serbian, German, French, and Spanish, the *Fiori di virtú* is an important link in the indirect transmission of our text.

It should be noted that, although *Physiologus* was one of the most popular handbooks of the Middle Ages, its legends and allegories were never received as "canonical" by later writers. Augustine for one was skeptical of the edifying legend of the pious pelican. In the very synthetic spirit of the work itself, the material was constantly manipulated to suit particular audiences. After narrating his version of the story of the frogs, the monk who composed the Byzantine recension of *Physiologus* draws the following startling observation in his allegory: "Thus are the majority of monks, who no sooner does a temptation grip them, but they allow themselves to become agitated with chattering and cry out, 'We have no rest in this monastery. Our prior does not take care of us, and our confreres scorn us. . . . Let us go off to another monastery where we might be bosses over others rather than the other way around.'"[45] As with most books which have made a difference in the way we think, *Physiologus* became all things to all men.

## THE TRANSLATION

The present translation is based on the two editions of the Latin *Physiologus* prepared by Francis Carmody (see Bibliography), the y- and the b-version. I have relied primarily on the y-version since it is generally agreed to be the closer of the two to the Greek original. Whenever important additions or variations are supplied by the b-version, however, I have translated them and put them in square brackets. It is astonishing but true that, outside the "Theobaldus" version, the Latin *Physiologus* has never been translated into English. Francis Carmody's translation for the Book Club of California is a "composite" text of the Greek, Syrian, Ethiopian, and Latin, while Carlill's is based on the Greek. T. H. White's *The Book of Beasts* is, of course, not *Physiologus* but a translation of the twelfth-century bestiary edited for the Roxburghe Club in 1928 by M. R. James. It appeared evident to me that a reliable translation of the Latin *Physiologus* was in order as this was the version common to Western Europe throughout the Middle Ages.

NOTES

1. The most comprehensive survey of *Physiologus's* general background and later influence remains F. Lauchert, *Geschichte des Physiologus* (Strassburg: Verlag Karl J. Trübner, 1889). Lauchert's study may be complemented by the studies of such later scholars as Florence McCulloch, Francis Klingender, Beryl Rowland, and Nikolaus Henkel (see Bibliography).
2. On the importance of paradoxography and medical-magical books in the transmission of *Physiologus's* legends, see the detailed study by Max Wellmann, "Der *Physiologos,* Eine religionsgeschichtlich-naturwissenschaftliche Untersuchung," *Philologus,* Supplementband 22, Heft 1 (1930):1–116.
3. One of the most popular versions of *Physiologus* was the "Theobaldus version" (late eleventh or early twelfth century), which was used as a school text sometimes provided with commentary during the later Middle Ages. See P. T. Eden, ed. and trans., *Theobaldi "Physiologus"* (Leiden: E. J. Brill, 1972). For a Waldensian redaction, see Alfons Mayer, "Der Waldensiche *Physiologus,*" *Romanische Forschungen* 5 (1890):392–418. On the use of animal legends in preaching, see G. R. Owst, *Literature and Pulpit in Medieval England* (Oxford: Basil Blackwell, 1966), pp. 196–204.
4. The Icelandic translation contains only the allegories. The absence of the legends might have been compensated for by the manuscript's rough drawings depicting the narrative matter. See Halldor Hermannson, *The Icelandic Physiologus, Facsimile Edition,* Icelandica, vol. 27 (Ithaca, N.Y.: Cornell University Press, 1938), esp. pp. 8–9.
5. Strabo (14. 1. 7) follows Aristotle in identifying the φυσιολόγοι with the pre-Socratics. In his description of Miletus, for example, he mentions Thales as the first to begin the study of φυσιολογία among the Greeks.

6. Diodorus Siculus 5. 40. 1–2.
7. Plutarch *On Isis and Osiris* 381B.
8. Ibid. 381B.
9. Ibid. 382B–C.
10. *Clemens Alexandrinus Zweiter Band Stromata, Buch I–VI*, herausgegeben . . . von Dr. Otto Stählin (Leipzig: J. C. Hinrichs, 1906), p. 249.
11. *Origenes Werke*, herausgegeben . . . von Prof. Dr. W. A. Baehrens (Leipzig: J. C. Hinrichs, 1925), p. 208. Origen also uses Paul's words, however, as a warning against idolatry of animal statues: "Ex hoc ergo inexcusabiles fiunt: quia cum manifestante Deo cognoverint Deum, non ut dignum est coluerunt, aut gratias egerunt; sed per cogitationum suarum vanitatem, dum formas et imagines requirunt in Deo, in semetipsis Dei imaginem perdiderunt, et qui jactare se in luce sapientiae videbantur, in profundas stultitiae tenebras devoluti sunt. Quid enim tam tetrum, tam obscurum, tamque tenebrosum, quam gloriam Dei ad corporalem et corruptibilem humanae formae effigiem vertere, ut moris est hiis qui simulacra venerantur, et divinae majestatis eminentiam volucribus, et quadrupedibus, et serpentibus exaequare?" (*Comment. in Epist. Ad Rom. Lib. I* [Migne, *PG* 14. 864]). On the other hand, the defense of animal symbolism as being opposed to idolatry is usually based among early Christian writers on Scriptures which often rely on such metaphoric language. Origen (*On the Wisdom of Solomon* 7. 17–21) observes: "Secundum haec intellige mihi etiam illud, quod ait: 'Naturas animalium irasque bestiarum'; nisi enim bene scisset 'naturas animalium,' numquam dixisset in evangeliis Salvator: 'dicite vulpi huic' neque Iohannes de quibusdam dixisset: 'serpentes generatio viperarum,' sed neque propheta diceret de nonnullis quod: 'equi admissarii facti sunt' et item alius: 'homo cum in honore esset, non intellexit; comparatus est iumentis insipientibus et similis factus est illis'" (Baehrens, ed., p. 211).
12. The analogy between the "scripture" of the natural

world and Holy Scripture occurs in Origen: "Ita
igitur cuncta secundum ea, quae prefati sumus, ex
visibilibus referri possunt ad invisibilia et a corpora-
libus ad incorporea et a 'manifestis' ad 'occulta,' ut
ipsa creatura mundi tali quadam dispensatione con-
dita intelligatur per divinam sapientiam, quae rebus
ipsis et exemplis invisibilia nos de visibilibus doceat
et a terrenis nos transferat ad caelestia. Haec autem
rationes non solum in creaturis omnibus habentur,
sed et ipsa scriptura divina tali quadam sapientiae
arte conscripta est." (Baehrens, ed., pp. 211–212).
For a discussion of this passage, see Jacques Chêne-
vert, *L'Église dans le Commentaire d'Origène sur le
Cantique des Cantiques*, Studia Travaux de recherche,
vol. 24 (Brussels: Desclée de Brouwer, 1969), pp.
88–91. Philo, the great Hellenized Jewish exegete,
lived ca. 13 B.C. to ca. A.D. 50 at Alexandria and
established the allegorical method of interpreting
Holy Scripture.

13. "Sed eleganter quidam sibi videntur physiologice
per allegoricam argumentationem de Saturno inter-
pretari, tempus esse, et ideo Coelum et Terram
parentes, ut et ipsos origini nullos, et ideo falcatum,
quia tempore omnia dirimantur, et ideo voratorem
suorum, quod omnia ex se edita in se ipsum con-
sumat" (Migne, *PL* 1. 674).

14. "Nam physiologia hoc habet, quia qui caret gaudio
sine dubio tristis est. Tristia autem vicina luctui est,
qui procreatur ex morte: unde haec esse apud in-
feros dicit" (*Servii Grammatici qui feruntur In Virgilii
Carmina Commentarii recenserunt Georgius Thilo et
Hermanus Hagen*, vol. 2 [Leipzig: B. G. Teubner,
1923], p. 53).

15. Rudolph Helm, ed., *Fabii Planciadis Fulgentii V. C.
Opera* (Stuttgart: B. G. Teubner, 1970), 92. 1, 5.

16. Francesco Sbordone, ed., *Physiologi Graeci singulas
variarum aetatum recensiones . . . in lucem protulit F.
Sbordone* (Milan: In Aedibus Societatis "Dante Ali-
ghieri-Alberighi, Segati et C.," 1936), p. 196.

17. See Emma Brunner-Traut, "Altägyptische Mythen

im *Physiologus*," *Antaios*, Band 10, 2 (July 1968):
184–198.

18. Wellmann, p. 13.

19. Fritz Hommel, ed., *Die aethiopische Übersetzung des Physiologus* (Leipzig: J. C. Hinrichs'sche Buchhandlung, 1877), p. xvi.

20. Lauchert, p. 65.

21. *Das "Buch der Naturgegenstände" herausgegeben und übersetzt von K. Ahrens* (Kiel, 1892), p. 17.

22. See my notes to chapter I on the lion.

23. Wellmann, pp. 10–11, esp. note 50.

24. Migne, *PL* 14. 261–262.

25. Lauchert, pp. 80–86.

26. Sbordone, p. lxxviii.

27. Lauchert, p. 89.

28. Ben E. Perry, "Physiologus," in *Pauly-Wissowa, Real-Encyclopädie der classischen Altertumswissenschaft*, vol. 20, pt. 1 (1941), p. 1120.

29. Lauchert, pp. 88–89.

30. See F. L. Cross and E. A. Livingstone, eds., *The Oxford Dictionary of the Christian Church*, 2d ed. (London: Oxford University Press, 1974), p. 385.

31. See esp. chaps. 4, 17, 20, 22, 34, and 36.

32. Wellmann, pp. 81–94.

33. Karl Ahrens, *Zur Geschichte des sogenannten Physiologus* (Ploen, 1885), pp. 2–3.

34. Herodotus *Hist.* 4. 109; Aesop *Fables* (ed. Halm) 189; Cicero *Oratio pro M. Aemilio Scauro*, frag. 4. 6–7; Apuleius *Metam.* 1. 9.

35. Pliny *Natural History* 8. 41, 59; Aelian *On Animals* 9. 16, 66.

36. Francesco Sbordone, *Ricerche sulle fonti e sulla composizione del Physiologus greco* (Naples: Arti Grafiche G. Torella & Figlio, 1936), p. 52; *Physiologi Graeci*, p. 37.

37. Plutarch *De sollertia animalium* 976D; Pliny *Natural History* 8. 62; 21. 39; Aelian *On Animals* 5. 40. This quality of the panther was a favorite theme among the illustrators of the bestiaries of the later Middle Ages. See Klingender, *Animals in Art and Thought*,

plates 216 and 219b; also, *Physiologus Bernensis, Voll-Faksimile-Ausgabe des Codex Bongarsianus 318 . . . von Christoph Steiger und Otto Hamburger* (Basel: Alkuin-Verlag, 1964), plate 17.

38. See Aristotle *History of the Animals* 9. 6. 612a12; Antigonos (ed. Westermann) 31 (37). 70; Pliny *Natural History* 8. 62; Aelian *On Animals* 5. 40, 8. 6 (ἡ πάρδαλις δὲ αἱρεῖ τῇ ὀσμῇ τὰ πλεῖστα, καὶ ἔτι μᾶλλον τὸν πίθηκον). It is very likely that the author subjected the legend of the panther to such drastic alteration in order to contrast it with the following chapter on the whale, which attracts fish with its sweet breath. Thus, the motif of the εὐωδία links a figure of damnation (the whale) with a corresponding figure of salvation (the panther).

39. Lauchert, pp. 79–80. The study of the later history of *Physiologus* has been advanced by the research of Lauchert, Henkel, McCulloch, and Perry (see Bibliography).

40. The *Liber Glossarum*, called the *Glossarium Ansileubi* by its early editors Pitra and Mai, was a collection of excerpts from the church fathers, Isidore, Hilarius, Eucherius, Fulgentius, Jumilius, Eutropius, and Orosius, as well as from Cicero's *Synonyma*, *Physiologus*, glosses on Vergil and Placidus, and Julian of Toledo; it was probably composed ca. 750 in Spain. See Max Manitius, *Geschichte der lateinischen Literatur des Mittelalters*, vol. 1 (Munich: C. H. Beck, 1965), pp. 133–134.

41. This important work has recently been edited and translated into English by P. T. Eden. See note 2.

42. See A. S. Cook, *The Old English Elene, Phoenix and Physiologus* (New Haven: Yale University Press, 1919), pp. 75–81.

43. McCulloch, p. 54.

44. See note 3.

45. Sbordone, *Ricerche*, p. 181.

## BIBLIOGRAPHY

The following bibliography is appended for those wishing further information about *Physiologus*.

Sbordone's edition (1936) is the starting point for any study of the Greek *Physiologus*; its massive textual apparatus and thorough listing of classical sources and Christian analogues allow it to supplant the other Greek texts from the first edition of Ponce de Leon (1587) down through those of Pitra (1855) and Lauchert (1889), although Kaimakis's recent edition (1974) shows that the last word has yet to be uttered. Carmody's b- and y-versions of the Latin *Physiologus* provide us with two important families of the Latin text. Both the *Dicta Chrysostomi* (ed. Heider, 1850) and the Theobaldus version (ed. Eden, 1972), though not taking us back to the original meaning, should be consulted where the reader's interest is focused on the later history of *Physiologus*. As has been noted already, the vernacular translations, particularly the Ethiopian (ed. Hommel, 1877) and the Syrian (ed. Land, 1874), are critical to any textual study of *Physiologus* since they antedate the surviving Greek manuscripts by many centuries.

Although not all of Lauchert's opinions about *Physiologus* are currently accepted, his *Geschichte des Physiologus* is in all respects a thorough and meticulous general study of the work, giving important information on the development of each legend, the origin, date, and sources, and the later history of the Greek and Latin text and vernacular translations. Lauchert also notes *Physiologus*'s influence on Christian art and edits a Greek and German version of *Physiologus*. To complement and update Lauchert's research, Wellmann's excellent study (1930) of the pseudoscientific literary background of *Physiologus* should be consulted, as well as Sbordone's *Ricerche* (1936), which corrects many inaccuracies in previous scholarship from the vantage point of a securely established textual tradition. The latest general survey of *Physiologus*'s history is Henkel's stimulating book (1976), which summarizes scholarship to date and provides much new research on the later influence of *Physiologus* on art, sculpture, and literature (particularly German literature) of the Middle Ages.

Students concerned with *Physiologus*'s place in art history will wish to consult Klingender (1971) for a general view of the subject and McCulloch (1962) for the bestiary tradition. The detailed study by Evans (1896) and Druce's essays are still worthwhile. Facsimile editions of two important manuscripts are to be found in James (1929) and Steiger/Hamburger (1964). T. H. White's translation (1954) and Beryl Rowland's charming book (1973) both provide numerous fine bestiary illustrations.

*Editions*

Cahier, Ch. *Nouveaux mélanges.* 4 vols. Paris, 1874–1877.

————, and Martin, A. *Mélanges d'archéologie, d'histoire et de littérature.* 4 vols. Paris, 1847–1856.

Carmody, Francis, ed. *Physiologus Latinus. Éditions préliminaires, versio B.* Paris: Librairie E. Droz, 1939.

————, ed. "Physiologus Latinus, versio Y." *The University of California Publications in Classical Philology* 12 (1941): 95–134.

Cook, A. S., ed. *The Old English Elene, Phoenix and Physiologus.* New Haven: Yale University Press, 1919.

Eden, P. T., ed. and trans. *Theobaldi "Physiologus."* Leiden: E. J. Brill, 1972.

Heider, G. "Dicta Joh. Crisostomi de naturis bestiarum." *Archiv für Kunde Österreicher Geschichts-Quellen* 5 (1850): 523–582.

Hermannson, Halldor. *The Icelandic Physiologus. Facsimile Edition.* Ithaca, New York: Cornell University Press, 1938.

Hommel, Fritz, ed. *Die aethiopische Übersetzung des Physiologus.* Leipzig: J. C. Hinrichs'sche Buchhandlung, 1877.

Kaimakis, Dimitris, ed. *Der Physiologus nach der ersten Redaktion*. Meisenheim am Glan: A. Hain, 1974.

Land, J. P. N. *Anecdota Syriaca*, Leiden, 1874. (Contains the Syrian translation of *Physiologus*.)

Lauchert, Friedrich. *Geschichte des Physiologus*. Strassburg: Verlag Karl J. Trübner, 1889. (A Greek text is found on pp. 229–279).

Mann, Max Friedrich. "Der Bestiaire Divin des Guillaume le Clerc." *Französische Studien* 6² (1888): 37–73.

Mauer, Friedrich, ed. *Der altdeutsche Physiologus, die Millstäter Reimfassung und die Wiener Prosa*. Tübingen: Max Niemeyer Verlag, 1967.

Mayer, Alfons. "Der waldensiche *Physiologus*." *Romanische Forschungen* 5 (1890): 392–418.

Pitra, J. B. *Spicilegium Solesmense*. Paris, 1855.

Ponce de Leon, G., ed. *S.P.N. Epiphanius ad Physiologum*. In *Patrologia Graeca*, edited by J. P. Migne, vol. 43, cols. 518–534. (This is a reprint of the edition of 1587, the earliest printed text of the Greek *Physiologus*, containing twenty-five chapters.)

Sbordone, F., ed. *Physiologus (Physiologi graeci singulas variarum aetatum recensiones . . . in lucem protulit F. Sbordone)*. Milan: In Aedibus Societatis "Dante Alighieri-Albrighi, Segati et C.," 1936.

Steiger, Christoph, and Hamburger, Otto, eds. *Physiologus Bernensis, Voll-Faksimile-Ausgabe des Codex Bongarsianus 318*. Basel: Alkuin-Verlag, 1964.

*Modern Translations*

Carlill, James. *Physiologus*. In *The Epic of the Beast*, edited by William Rose. New York: E. P. Dutton & Co., 1924.

Carmody, Francis. *Physiologus, The Very Ancient Book of Beasts, Plants and Stones*. San Francisco: The Book Club of California, 1953.

Cook, A. S. *The Old English Physiologus*. Yale Studies in English, vol. 63. New Haven: Yale University Press, 1921.

Eden, P. T. (see above).

Elliott, Thomas J. *A Medieval Bestiary*. Boston: Godine, 1971.

Rendell, Alan Wood. *Physiologus. A Metrical Bestiary of Twelve Chapters by Bishop Theobald, Printed in Cologne, 1492*. London: J. & E. Bumpus, 1928.

Seel, Otto. *Der Physiologus, Übertragen und Erläutert von Otto Seel*. Zurich: Artemis Verlag, 1960.

White, T. H. *The Bestiary, A Book of Beasts*. New York: G. P. Putnam's Sons, 1954.

*General Studies*

Broek, R. van den. *The Myth of the Phoenix according to Classical and Early Christian Traditions*. Leiden: E. J. Brill, 1972.

Carmody, Francis. *"De Bestiis et Aliis Rebus* and the Latin *Physiologus." Speculum* 13 (April 1938):153–159.

Cronin, Grover, Jr. "The Bestiary and the Medieval Mind—Some Complexities." *Modern Language Quarterly* 2 (1941):191–198.

Druce, G. C. "An Account of the μυρμηκολέων or Antlion." *The Antiquaries Journal* 3 (1923):47–64.

———. "The Elephant in Medieval Legend and Art." *Archeological Journal* 76 (1919):1–73.

———. "The Medieval Bestiaries and Their Influence on Ecclesiastical Decorative Art." *The Journal of the British Archeological Association*, n.s. 25 (1919):41–82.

Evans, E. P. *Animal Symbolism in Ecclesiastical Architecture*. London: W. Heinemann, 1896.

Gervaise. "Le Bestiaire de Gervaise." Edited by Paul Meyer. *Romania* 1 (1872):420–443.

Goldstaub, Max. "Der Physiologus und seine Weiterbildung besonders in der lateinischen und in der Byzantinischen Literatur." *Philologus*, Supplementband 8, Heft 3 (1899–1901):339–404.

Guillaume le Clerc. *Das Thierbuch des Normanischen Dichters Guillaume le Clerc*. Edited by Robert Reinsch. Leipzig, 1892.

Henkel, Nikolaus. *Studien zum Physiologus im Mittelalter*. Tübingen: Max Niemeyer Verlag, 1976.

James, M. R., ed. *The Bestiary, Being a Reproduction in Full of the Manuscript Ii.4.26 in the University Library,*

*Cambridge*. Oxford: J. Johnson, at the University Press, 1928.

Klingender, Francis. *Animals in Art and Thought to the End of the Middle Ages*. Edited by Evelyn Antal and John Harthan. Cambridge, Massachusetts; MIT Press, 1971.

Lauchert, Friedrich. *Geschichte des Physiologus*. Strassburg: Verlag Karl J. Trübner, 1889.

McCulloch, Florence. *Medieval Latin and French Bestiaries*. University of North Carolina Studies in Romance Languages and Literatures, no. 33. Chapel Hill, N.C.: University of North Carolina Press, 1962.

Perry, Ben E. "Physiologus." *Pauly-Wissowa, Real-Encyclopädie der classischen Altertumswissenschaft*, vol. 20, pt. 1 (Stuttgart: J. B. Metzlersche Verlagsbuchhandlung, 1941), cols. 1074–1129.

Robin, P. Ansell. *Animal Lore in English Literature*. London: John Murray, 1932.

Rowland, Beryl. *Animals with Human Faces*. Knoxville, Tenn.: University of Tennessee Press, 1973.

Sbordone, Francesco, *Ricerche sulle fonti e sulla composizione del Physiologus greco*. Naples: Arti Grafiche G. Torella & Figlio, 1936.

Philippe de Thaon. *Le Bestiaire de Philippe de Thaün*. Edited by Emmanuel Walberg. Lund: H. Möller, 1900.

Treu, Ursula. "Bestiary." *Encyclopedia Britannica*, vol. 3. 1970.

Wellmann, Max. "Der *Physiologos*, Eine religionsgeschichtlich-naturwissenschaftliche Untersuchung." *Philologus*, Supplementband 22, Heft 1 (1930):1–116.

White, Beatrice. "Medieval Animal Lore." *Anglia* 72 (1954):21–30.

White, Lynn, Jr. "Natural Science and Naturalistic Art in the Middle Ages." *American Historical Review* 52 (1947):421–435.

NOTE TO THE PAPERBACK EDITION

The reprinting of my book thirty years after it first appeared is a tribute to the richness of the tradition to which *Physiologus* belongs and the diversity of scholarly approaches that it encourages.

A new complexity entered into the history of *Physiologus* with the appearance of the earliest surviving illustrations to the text in the ninth-century Bern manuscript (*Burgerbibliothek Codex Bongarsianus 318*). The interaction between text and image, particularly in the amply illustrated manuscripts of the bestiary tradition of the later Middle Ages, has been a topic of keen interest for over a century. Scholars interested in recent discussions of this interaction, with all its complexities and contradictions, can turn to publications such as Janetta Rebold Benton's *Medieval Menagerie: Animals in the Art of the Middle Ages* (New York, 1992) and Debra Hassig's *Medieval Bestiaries: Text, Image, Ideology* (Cambridge, 1995). Hassig takes a semiotic approach to the subject in an effort to link the bestiary content to contemporary medieval life and to show the multiplicity of interpretations that readers brought to the text. She studies twenty-eight English manuscripts from the twelfth through the fourteenth century. Both Benton's and Hassig's books are richly illustrated.

Interdisciplinary approaches that examine the literary, historical, social, and political uses to which *Physiologus* and the bestiary were summoned can be found in *Beasts and Birds of the Middle Ages: The Bestiary and Its Legacy*, edited by Willene B. Clarke and Meradith T. McMunn (Philadelphia, 1989), and in *The Mark of the Beast: The Medieval Bestiary in Art, Life, and Literature* (New York, 1999), a collection of essays edited by Debra Hassig. Similarly, Malcolm South's *Mythical and Fabulous Creatures* (Westport, Connecticut, 1987) is a useful reference guide to a wide variety of fabulous creatures—many of which made their debut in

*Physiologus*—and their deployment in art, literature, and history. Important for explaining the heightened interest in the bestiary tradition during the High Middle Ages is Joyce E. Salisbury's book *The Beast Within: Animals in the Middle Ages* (New York, 1994), which argues that a paradigm shift in Western attitudes toward animals occurred in the twelfth century, and marked a breakdown in the rigid conceptual separation of mankind and animals.

The appearance of medieval vernacular translations of *Physiologus* signaled an important stage in the transmission of the text. Some recent editions of the English versions can be found in *The Old English Physiologus,* edited by Ann Squires (Durham, England, 1988), and in *The Middle English Physiologus,* edited for the Early English Text Society by Hanneke Wirtjes (Oxford, 1991), the latter of which presents a new edition of *BL MS Arundel 292,* the only surviving manuscript of this Middle English poetic version of *Physiologus.* For the Old High German version, see Christian Schröder's *Der Millstätter Physiologus: Text, Übersetzung, Kommentar* (Würzburg, 2005).

The influence of *Physiologus* on the ancient genre of the animal fable is well established and can be seen in *The Fables of Odo of Cheriton* (Syracuse, NY, 1985), translated and with an introduction by John C. Jacobs. Jacobs's translation is accompanied by drawings of animals from the pen of the thirteenth-century architect Villard de Honnecourt. Jeanette Beer's translation of Richard de Fournival's bestiary (*Master Richard's Bestiary of Love and Response* [Berkeley, 1986]), one of the most popular of all French bestiaries, was followed by her recent study (*Beasts of Love: Richard de Fournival's "Bestiaire d'amour" and a Woman's Response* [Toronto, 2003]), an analysis of the rhetoric or casuistry of love that Richard grafted onto traditional bestiary lore. Another well-known Old French prose bestiary of the early thirteenth century has been translated by Guy R. Mermier (*A Medieval Book of Beasts: Pierre de Beauvais' Bestiary* [Lewiston, NY, 1992]).

Scholars with an interest in the Greek versions of *Physiologus* and the traditions it inspired will want to consult recent work on the Greek texts, particularly *Physiologus: The Greek and Armenian Versions with a Study of Translation Technique,* by Gohar Muradyan (Leuven, Belgium, 2005), and *Physiologos: Le bestiaire des bestiaires: texte traduit du grec, introduit et commenté,* by Arnaud Zucker (Grenoble, 2004).

Finally, Simona Cohen's recent discussion of the bestiary (*Animals as Disguised Symbols in Renaissance Art* [Leiden, Netherlands, 2008]) has opened a new chapter in the discussion of the topic by demonstrating the continued vitality of the bestiary well into the Renaissance.

Michael J. Curley
Tacoma, Washington
2009

# PHYSIOLOGUS

### 1. We begin first of all by speaking of the Lion, the king of all the beasts

Jacob, blessing his son Judah, said, "Judah is a lion's whelp" [Gen. 49:9]. Physiologus, who wrote about the nature of these words, said that the lion has three natures. His first nature is that when he walks following a scent in the mountains, and the odor of a hunter reaches him, he covers his tracks with his tail wherever he has walked so that the hunter may not follow them and find his den and capture him. Thus also, our Savior, the spiritual lion of the tribe of Judah, the root of David [cf. Rev. 5:5], having been sent down by his coeternal Father, hid his intelligible tracks (that is, his divine nature) from the unbelieving Jews: an angel with angels, an archangel with archangels, a throne with thrones, a

4

power with powers, descending until he had descended into the womb of a virgin to save the human race which had perished. "And the word was made flesh and dwelt among us" [John 1:14]. And those who are on high not knowing him as he descended and ascended said this, "Who is this king of glory?" And the angels leading him down answered, "He is the lord of virtues, the king of glory" [cf. Ps. 24:10].

The second nature of the lion is that, although he has fallen asleep, his eyes keep watch for him, for they remain open. In the Song of Songs the betrothed bears witness, saying, "I sleep, but my heart is awake" [S. of S. 5:2]. And indeed, my Lord physically slept on the cross, but his divine nature always keeps watch in the right hand of the Father [cf. Matt. 26:64]. "He who guards Israel will neither slumber nor sleep" [Ps. 121:4].

The third nature of the lion is that, when the lioness has given birth to her whelp, she brings it forth dead. And she guards it for three days until its sire arrives on the third day and, breathing into its face on the third day, he awakens it. Thus did the almighty Father of all awaken from the dead on the third day the firstborn of every creature [cf. Col. 1:15]. Jacob, therefore, spoke well, "Judah is a lion's whelp; who has awakened him?" [Gen. 49:9].

## II. *On the Antelope*

There is an animal called the antelope, so exceedingly alert that the hunter is unable to approach him. He has long horns in the shape of a saw so that he is able to cut through thick, high trees and fell them to the ground. If he is thirsty, however, he goes to the terrible Euphrates River and drinks. Near the river there are

herecine (as they are called in Greek), that is, shrubs
with thin branches; and he comes up to that herecine
shrub frisking about and is ensnared in its branches.
Then he cries out, wishing to escape, and is unable to do
so for he has been ensnared. Hearing him, the hunter
will come and slay him.

O citizen whose commonwealth is heavenly [Phil
3:20], confiding in the two horns of might abstain from
acts of slander and pleasure, inimical desire and the
pomp of the world; the powers of the angels will rejoice for you, for the two horns are the two Testaments.
But watch that you are not held by this most delicate
herecine, that is, by the little shrub which covers you at
just the right moment. Watch that you are not held in its
snare, for the wicked hunter (that is, the devil) will slay
you. The wise man, however, flees wine and women
[cf. Ecclesiasticus 19:2 and Hos. 4:11].

### III. *On Piroboli Rocks*

Piroboli rocks exist only in the East; they are igneous
rocks of masculine and feminine gender. As long as
they are separate from one another, they do not burn,
but, if the male approaches the female, fire breaks forth
and consumes all.

O citizen moderate in all things, many are the men in
Tartarus who fell into temptations on account of women
[cf. Ecclesiasticus 9:8]. [Separate yourself far from
women lest the double fire be ignited when you approach one another and consume the good things which
Christ has bequeathed you. Satan's angels always battle
the just, not only holy men but also chaste women.
Samson and Joseph were both tempted by women; one
conquered and the other was himself conquered. Eve
and Susanna were tempted: the former conquered and
the latter was conquered.]

### IV. *On the Swordfish*

Physiologus spoke well of those who abstain from all
things but who do not persevere to the end [cf. Matt.
24:13]. There is an animal in the sea, he said, called the
swordfish, which has long wings; and, when he sees
the ships sailing, he imitates them and raises his wings
and strives with the ships as they sail. Growing tired
after racing three or four miles or more, he folds up his
wings and the waves carry him back to his former abode
where he was at first.

The sea is the world, the ships are the prophets and
apostles who cross through this world, [through the eye
of the squall and storm of this world without any danger
or shipwreck to their faith; they conquer the deadly
waves, that is, the contrary] powers of the adversary.

The swordfish who does not keep pace with the cross-
ing ships represents those who are abstinent for a time
but who do not persevere with good pace. These begin
with good works but do not persevere to the end because of greed, pride, and love of wicked gain [cf. Tit.
1:11] or acts of fornication or adultery or hatred; within
these waves of the sea (that is, the contrary powers)
are the powers which carry them down into hell.

### v. *On the Charadrius*

There is another kind of flying animal called the charadrius mentioned in Deuteronomy [Deut. 14:18] which is
entirely white with no black part at all. His excrement is
a cure for those whose eyes are growing dim and he is
found in the halls of kings. If someone is ill, whether he
will live or die can be known from the charadrius. The

bird turns his face away from the man whose illness will
bring death and thus everyone knows that he is going to
die. On the other hand, if the disease is not fatal, the
charadrius stares the sick man in the face and the sick
man stares back at the charadrius, who releases him
from his illness. Then, flying up to the atmosphere of
the sun, the charadrius burns away the sick man's illness and scatters it abroad.

The charadrius is our Savior and the one who is ill
receives the good person of our Savior who is entirely
white without spot or wrinkle [cf. Eph. 5:27]. The Lord
himself said in the Gospel, "The ruler of this world is
coming, and he will find nothing against me" [John 14:
30]. For "He committed no sin; no guile was found on
his lips" [I Peter 2:22]. And coming down from heaven
to the Jewish people, he turned his divine nature from
them, saying, "Behold, your abandoned house is forsaken" [Lk. 13:35]. Yet he came down to us, the Gentiles, taking away our infirmities [Matt. 8:17] and carrying off our sins [cf. Is. 53:4]; and he was lifted up to the
wood of the cross: "Ascending on high, he led a host of
captives, and he gave gifts to men" [Ps. 68:18 and Eph.
4:8]. ["Indeed, those who believed not, received not.
But to all who received him, who believed in his name,
he gave power to become children of God" [John 1:12].]

But you say that the charadrius is unclean according
to the law, therefore how can he represent the person of
the Savior? The serpent is unclean, yet the Lord himself
bore witness concerning him in the Gospel, saying,
"And as Moses lifted up the serpent in the wilderness,
so must the Son of Man be lifted up" [John 3:14]. The
serpent was called a rather wise beast, as were the lion
and many others. The creatures are twofold: the praiseworthy and the blameworthy. [Similarly, both the lion
and the eagle are unclean even though the one is the
king of the beasts and the other of the birds. Because of
their kingdoms they are likened to Christ, yet because of

their greediness they are likened to the devil. And there are many others among the creatures who have double significances; certain are praiseworthy while others are blameworthy, according to their different habits and nature.]

### VI. *On the Pelican*

David says in Psalm 101, "I am like the pelican in loneliness" [Ps. 102:7]. Physiologus says of the pelican that it is an exceeding lover of its young. If the pelican brings forth young and the little ones grow, they take to striking their parents in the face. The parents, however, hitting back kill their young ones and then, moved by compassion, they weep over them for three days, lamenting over those whom they killed. On the third day, their mother strikes her side and spills her own blood over their dead bodies (that is, of the chicks) and the blood itself awakens them from death.

Thus did our Lord speaking through Isaiah say, "I have brought forth sons and have lifted them up, but they have scorned me" [Is. 1:2]. The Maker of every creature brought us forth and we struck him. How did we strike him? Because we served the creature rather than the creator [cf. Rom. 1:25]. The Lord ascended the height of the cross and the impious ones struck his side and opened it and blood and water came forth for eternal life [cf. John 19:34 and 6:55], blood because it is said, "Taking the chalice he gave thanks" [Matt. 26:27 and Lk. 22:17], and water because of the baptism of repentance [Mk. 1:4 and Lk. 3:3]. The Lord said, "They have forsaken me, the fountain of living water," and so on [Jer. 2:13]. Physiologus, therefore, spoke well of the pelican.

## vii. *On the Owl*

D avid says of the owl in the same Psalm 101, "I have become like an owl in the house" [Ps. 102:6]. The owl is this kind of bird: he loves the darkness more than the light.

Our Lord Jesus Christ loved those who were in the darkness and the shadow of death [cf. Is. 9:2], the Gentiles and the Jews who then received adoption as sons [cf. Gal. 4:5] and the promise of the patriarchs [cf. Rom. 15:8]. Concerning this the Savior said, "Fear not little flock for it is your Father's pleasure to give you the kingdom" [Lk. 12:32]. And the Prophet says, "Those who were not my people I will call 'my people,' and her who was not beloved I will call 'my beloved'" [Rom. 9:25

and Hos. 2:23]. Yet, following Deuteronomy, you say
that the owl is an unclean bird. But the Apostle said,
"For our sake he made him to be sin who knew no sin"
[II Cor. 5:21] and "He became all things to all men, that
he might provide all profit" [cf. I Cor. 9:22].

[This beast is the figure of the Jewish people who,
when our Lord and Savior came to save them, rejected
him, saying, "We have no king but Caesar, we know
not who this man is" [John 19:15]. Thus, they loved the
darkness more than the light. Then the Lord turned to
us Gentiles and illuminated us while we were sitting in
the darkness and in the region of the shadow of death,
and in the region of the shadow of death light rose up
for us [cf. Is. 9:2]. The Savior spoke through the Prophet about this people, saying, "People whom I knew not
served me" [Ps. 18:43]. And, elsewhere, "Those who
were not my people I will call 'my people,' and her who
was not beloved I will call 'my beloved'" [Rom. 9:25
and Hos. 2:23]. Concerning the Jewish people who pre-

ferred the darkness to the light [cf. John 3:19], the Lord says in the psalm, "Alien sons have lied to me, and they have grown old, and stumbled from their paths" [Ps. 18:44–45].]

## VIII. *On the Eagle*

David says in Psalm 102, "Your youth will be renewed like the eagle's" [Ps. 103:5]. Physiologus says of the eagle that, when he grows old, his wings grow heavy and his eyes grow dim. What does he do then? He seeks out a fountain and then flies up into the atmosphere of the sun, and he burns away his wings and the dimness of his eyes, and descends into the fountain and bathes himself three times and is restored and made new again.

Therefore, you also, if you have the old clothing and the eyes of your heart have grown dim, seek out the spiritual fountain who is the Lord. "They have forsaken me, the fountain of living water" [Jer. 2:13]. As you fly into the height of the sun of justice [Mal. 4:2], who is Christ as the Apostle says, he himself will burn off your old clothing which is the devil's. Therefore, those two elders in Daniel heard, "You have grown old in wicked days" [Dan. 13:52]. Be baptized in the everlasting fountain, putting off the old man and his actions and putting on the new, you who have been created after the likeness of God [cf. Eph. 4:24] as the Apostle said. Therefore, David said, "Your youth will be renewed like the eagle's" [Ps. 103:5].

## IX. *On the Phoenix*

The Savior said in the Gospel, "I have the power to lay down my life, and I have the power to take it again" [John 10:18]. And the Jews were angered by his words. There is a species of bird in the land of India which is called the phoenix, which enters the wood of Lebanon after five hundred years and bathes his two wings in the fragrance. He then signals to the priest of Heliopolis (that is, the city named Heliopolis) during the new month, that is, Adar, which in Greek is called Farmuti or Phamenoth. When the priest has been signaled, he goes in to the altar and heaps it with brushwood. Then the bird enters Heliopolis laden with fragrance and mounts the altar, where he ignites the fire and burns himself up. The next day then the priest examines the altar and finds a worm in the ashes. On the second day, however, he finds a tiny birdling. On the third day he finds a huge eagle which taking flight greets the priest and goes off to his former abode.

If this species of bird has the power to kill himself in such a manner as to raise himself up, how foolish are those men who grow angry at the words of the Savior, "I have the power to lay down my life, and I have the power to take it again" [John 10:18]. The phoenix represents the person of the Savior since, descending from the heavens, he left his two wings full of good odors (that is, his best words) so that we, holding forth the labors of our hand, might return the pleasant spiritual odor to him in good works. Physiologus, therefore, speaks well of the phoenix.

## x. *On the Hoopoe*

The law says, "Whoever curses his father or mother will die the death" [Ex. 21:17]. What then of the patricide and the matricide? There is a bird called the hoopoe; if

the young see their parents grow old and their eyes dim,
they preen the parents' feathers and lick their eyes and
warm their parents beneath their wings and nourish
them as a reciprocation just as they nourished their
chicks, and they become new parents of their own
parents. [Their feathers are renewed and their eyes re-
illuminated so that they are able to be utterly renewed
and to see and fly as before. Afterward, they thank their
young who piously rendered them service.] And after a
fashion they say to their parents, "Just as you labored
to nourish us, so we also do likewise for you."

How unreasonable are these men who do not love
their parents! Physiologus, therefore, spoke well of the
hoopoe.

## XI. *On the Wild Ass*

It is said in Job, "Who has let the wild ass go free?"
[Job 39:5]. Physiologus says of the wild ass that being
first among those of the herd, if he begets colts, the
father will break their necessaries so that they produce
no seed.

The patriarchs tried to create carnal seed, but the
apostles spiritually obtained carnal sons so that they
might possess heavenly seed, just as it is said, "Rejoice,
barren woman who does not bear," and so on [Is. 54:1].
The Old Testament announces the seed, but the New
proclaims abstinence.

## XII. *On the Viper*

John said to the Pharisees, "You brood of vipers" [Matt.
3:7 and Lk. 3:7]. Physiologus says of the viper that the
male has the face of a man, while the female has the

16

form of a woman down to her navel, but from her navel down to her tail she has the form of a crocodile. Indeed, the woman has no secret place, that is, genitals for giving birth, but has only a pinhole. If the male lies with the female and spills his seed into her mouth, and if she drinks his seed, she will cut off the male's necessaries (that is, his male organs) and he will die. When, however, the young have grown within the womb of their mother who has no genitals for giving birth, they pierce through her side, killing her in their escape.

Our Savior, therefore, likened the Pharisees to the viper; just as the viper's brood kills its father and mother, so this people which is without God kills its father, Jesus Christ, and its earthly mother, Jerusalem. "Yet how will they flee from the wrath to come?" [Lk. 3:7]. Our father Jesus Christ and mother church live in eternity while those living in sin are dead.


XIII. *On the Serpent*

The Savior says, "Be wise like the serpents and mild like the doves" [Matt. 10:16]. The serpent has three natures. The first nature is this: when he grows old, his eyes become dim and, if he wants to become new again, he abstains and fasts for forty days until his skin becomes loosened from his flesh. And if it does become loosened with fasting, he goes and finds a narrow crack in the rock, and entering it he bruises himself and contracts and throws off his old skin and becomes new again. We, too, throw off for Christ the old man and his clothing through much abstinence and tribulation. And you, seek out Christ the spiritual rock and the narrow crack. "The gate is narrow and there is tribulation on the way which leads toward life, and few are those who enter through it" [Matt. 7:14].

His second nature is this: when he comes to the river to drink water, he does not bring with him the poison which he bears in his head, but he leaves it in his pit. Therefore, when we gather together we, too, ought to draw the living and everlasting water [cf. John 4:15]. And when we hear the divine and heavenly word in church, we ought not to bear poison along with us (that is, wicked earthly desires) [cf. Col. 3:5]. Many are the fools who did not come to that spiritual banquet; some bought five yolk of oxen, some sought a field, others took a wife [cf. Lk. 14:16–24]. As the Gospel states, "Pay all of them, therefore, their dues, fear to whom fear is due, honor to whom honor is due, taxes to whom taxes are due," and so on [Rom. 13:7].

The third nature of the serpent is this: when he sees a naked man he fears him, but when he sees a man clothed he attacks him. Spiritually we, too, ought to understand that when the first man, our father Adam, was naked in paradise, the serpent did not succeed in attacking him, but when he dressed in a tunic (that is, the mortality of a sinful fleshly body), then the serpent assaulted him. Therefore, if you also have mortal dress, that is, the old man, and you want to hear the words, "You have grown old in wicked days" [Dan. 13:52], the serpent will assault you. If you deprive yourself, however, of the clothing of princes and powers of this world, of the rulers and spirits of wickedness in heavenly places as the Apostle said [cf. Eph. 6:12], then the serpent will not be able to attack you. Physiologus spoke wisely, therefore, of the serpent.

The fourth nature of the serpent is this: when a man approaches seeking to kill him, the serpent surrenders his entire body to the blows but protects his head. In time of temptation we, too, ought to surrender our entire body but protect our head, that is, we ought not to deny Christ. All martyrs acted in this fashion, "For the head of every man is Christ" [I Cor. 11:3].

## xiv. *On the Ant*

Solomon said in Proverbs, "Go to the ant, O sluggard,
and consider her ways" [Prov. 6:6]. Physiologus said
that the ant has three natures. The first nature is this:
ants walk in order, each one carrying a grain in his
mouth. The ants who have nothing do not say to the
others, "Give us your grain" [cf. Matt. 25:8], but they
pass over the tracks of the others and reach a place
where they find the grain; taking it up, they carry it off
to their dwelling. This story is told about imprudent and
unreasonable people. Woe to those virgins who be-
seeched the wise ones, saying, "Give us oil from your
lamps, since ours are going out" [Matt. 25:8]. The others, however, heard them and being reasonable and in-
telligent said, "We cannot, for perhaps there will not be
enough for us and for you" [Matt. 25:8].

[These things have been spoken about irrational animals and weak reptiles since they behave so prudently
that none of them is foolish but all are found to be clever
and wise. How much more ought those five rational virgins who were made foolish through their own negli-
gence imitate the five wise ones and get oil for their own
lamps from wherever those others got it. They ought
not to have asked the wise ones through their own idle-
ness and foolishness, saying, "Give us oil from your
lamps" [Matt. 25:8]. O what empty foolishness! If they
were not able to find oil on their own wherever the
others found it, they should have imitated the inge-
niousness of the ants. While the foolish virgins expected
oil from others' lamps, the Bridegroom came and they
remained outside with extinguished lamps.]

The second nature of the ant is this: when it has hidden the grain in its dwelling, it separates it into two
parts so that winter might not destroy it nor the flooding
rains germinate it and the ant perish of hunger. And
you, separate the words of the Old Testament, the carnal from the spiritual, lest the letter kill you when it
germinates. [And you, man of God, divide the Old Testament scriptures into two parts, that is, according to
the story and its spiritual meaning. Divide truth from
fiction, separate the spiritual from the corporeal, tran-
scend the killing letters toward the life-giving spirit,
lest while the letter is germinating on a winter's day
(that is on the Day of Judgment) you die of hunger.]
Paul the Apostle says, "The law is spiritual" [Rom. 7:
14]. And later he says, "The letter kills, but the spirit gives
life" [II Cor. 3:6]. And later, "Those things which give
life are the two Testaments" [cf. Gal. 4:24]. The Jews,
however, regarding the letter alone perished of hunger
and became murderers of the prophets and of God,
peeling the rods that the flocks might give birth [cf.
Gen. 30:37f.], having carnal circumcision, sabbaths,

and feasts of the tabernacles. But all of these are spiritual and intelligible things. [And you, O man of God,
peel the rods and expose the white as Jacob did. Throw
them into the water so that your flocks will bear clean
and spiritual fruit and not produce carnal and corrupt
offspring.]

The third nature of the ant is this: at harvest time, he
goes into the fields, climbs up the ears, and bears away
the grain. But, before climbing up the ears, the ant
catches their scent from beneath and perceives from the
scent whether it is wheat or barley. If it is barley, he
immediately rushes off to the ear of wheat since barley
is the food of brutes. And Job says, "Let barley come
forth for me instead of wheat" [Job 31:40]. [Flee, O man
of God, barley, that is the teachings of the heretics, for
they are barley and harmful things fit to be cast among

rocks. Heresies kill the souls of men. Flee Sabellius, Marcion, Manichaeus, avoid Novatianus, Montanus, Valentinus, Basilides, Macedonius, avoid Donatianus and Photinus and all who come forth from the Arian brood like serpentine offspring from the womb of the dragon. The dogmas of these men are false and hostile to truth.] And the Prophet says, "Flee Babylon and flee from the land of the Chaldeans" [Jer. 50:8], that is, flee alien teaching of alien glory; it is like barley-food, it kills the soul (for it is said to be and is an enemy of truth). The story of the ant was wisely spoken.

## xv. *On the Siren and Ass-Centaur*

Formerly, Isaiah the Prophet pointed out that the sirens and ass-centaurs and hedgehogs will come into Babylon and dance [cf. Is. 13:21 and 34:14]. Physiologus treated the nature of each one, saying of the sirens that they are deadly animals living in the sea which cry out with odd voices, for the half of them down to the navel bears the figure of a man, while the other half is that of a bird. [They sing a most pleasing song so that through the sweetness of the voice they charm the hearing of men sailing far away and draw them to themselves. By the great sweetness of their extended song they charm the ears and senses of the sailors and put them to sleep. When they see the men lulled by most heavy sleep, they attack them and tear them to pieces. Thus, they deceive men unacquainted with the persuasion of their voices and kill them. Just so are those men deceived who delight in the charms of the world, in games and the pleasures of the theater. Dissipated by tragedies and various melodies and lulled to sleep, these men become the prey of their enemies.]

Likewise, the ass-centaurs from their breasts up bear the figure of a man and that of an ass from there down. "Thus the man of deceitful heart is confused in all his ways" [Jas. 1:8]. Such are the impulses of the souls of wicked merchants; they even sin secretly while gathered together in church. As the Apostle said, "Holding the form of piety, they deny its virtue" [II Tim. 3:5]. And in church their souls are like sheep, yet when they are released from the congregation they become like the herd. "They are like brutish beasts" [Ps. 49:20].

Such beasts, sirens or ass-centaurs, represent the figures of devils.

## XVI. *On the Hedgehog*

The hedgehog does not quite have the appearance of a ball as he is full of quills. Physiologus said of the hedgehog that he climbs up to the grape on the vine and then throws down the berries (that is, the grapes) onto the ground. Then he rolls himself over on them, fastening the fruit of the vine to his quills, and carries it off to his young and discards the plucked stalk.

And you, O Christian, refrain from busying yourself about everything and stand watch over your spiritual vineyard from which you stock your spiritual cellar. Make a cache in the halls of God the King, in the holy tribunal of Christ, and you will receive eternal life. [Do not let concern for this world and the pleasure of temporal goods preoccupy you, for then the prickly devil, scattering all your spiritual fruits, will pierce them with his quills and make you food for the beasts. Your soul will become bare, empty and barren like a tendril without fruit. After this you will cry out, "My own vineyard I have not kept" [S. of S. 1:6], as the scripture of the Song of Songs bears witness.]

In such a way have you allowed that most wicked spirit to climb up to your place, and he has scattered your abstinence. Thus he has deceived you with the barbs of death in order to divide your plunder among hostile powers. Rightly, therefore, did Physiologus compare the ways of animals to spiritual matters.

### XVII. *On the Ibis*

There is an animal known as the ibis which according to the law is unclean beyond all other birds [cf. Lev. 11:17] [since it always feeds on carrion along the shore of the sea or rivers or swamps, where it walks by day and night seeking dead fish or carrion cast up putrid and decayed from the water]. Not knowing how to swim, the ibis feeds along the banks of rivers or ponds. He cannot swim in the depths but only where unclean little fish dwell, and is found outside the most deep places.

Learn how to swim spiritually so that you may come into the deep river, intelligible and spiritual, and to the depth of the wisdom of the power of God [cf. Rom. 11:33]. If you want to go into the deep, and to learn the mysteries of Jesus Christ, learn to swim spiritually. [Take the spiritual and most clean food which the Apostle names, saying, "The fruit of the spirit is love, joy, peace, patience, long-suffering, goodness, kindness, meekness, faith, modesty, continence, and chastity" [Gal. 5:22–23]. Yet by avoiding the deep water where you might get spiritual food, and going along the shore and wandering, you will be fed on dead and stinking carrion. The Apostle said of these, "Now the works of the flesh are plain: impurity, adultery, fornication, immodesty, lust, idolatry, drunkenness, avarice, and covetousness" [Gal. 5:19–21]. These are the carnal and deadly foods by which unhappy souls are nourished to suffering.]

Unless you extend your two hands and make the figure of the cross, you will not be able to pass through the sea. And, unless you want to pass through the world to God through the cross, you will not avoid scandals. Wise men who do not know how to swim or to pray, feed outside the Church. Outside the faith are deeds of fornication, adultery, slander, and greed. The root of all evils is greed [I Tim. 6:10], and the figure of the cross is over every creature.

And the sun is unable to shine unless it extends its rays. And the moon does not shine unless it extends its two horns. And even the birds of the sky are unable to fly unless they extend their wings. Also, unless its sail is extended, a ship will not be set in motion for sailing when the winds blow. So also Moses, extending his hands, killed Amalech [cf. Ex. 17:11]. Because Daniel prayed, he evaded the lions [Dan. 6:19–23]. Similarly, Jonah was saved by praying when he was in the belly of the whale [Jon. 2]. Thecla was thrown into the fire and into the pits of the beasts and the figure of the cross saved her. Susanna, too, was set free from the elders. Judith killed Holofernes thus with her mighty right hand, and Esther killed Haman [Est. 7]. The three boys in the fiery furnace were protected by such a sign [Dan. 3:1–30]. Many holy men endured other worse kinds of torments. [All holy men are represented by this figure. Like birds migrating, they arrive in the celestial kingdom as in a most calm port. Those not knowing how to swim spiritually and who dally with earthly and mortal things will be shut out of the celestial kingdom and perish among the dead. Thus, the Lord said in the Gospel, "Leave the dead to bury their own dead" [Matt. 8:22].] Physiologus, therefore, spoke wisely of the ibis.

## xviii. *On the Fox*

The fox is an entirely deceitful animal who plays tricks. If he is hungry and finds nothing to eat, he seeks out a rubbish pit [where there is red earth and rolls in it so that he appears bloodied all over; he then throws himself down and rolls over as though dead]. Then, throwing himself on his back, he stares upwards, draws in his breath, and thoroughly bloats himself up. Now the birds, thinking the fox dead, descend upon him to devour him. But he stretches out and seizes them, and the birds themselves die a miserable death.

[The fox is a figure of the devil. To those who live according to the flesh he pretends to be dead. Although he may hold sinners within his gullet, to spiritual men and those perfected in faith, however, he is dead and reduced to nothing.] The devil is, in fact, utterly dead as is the effect of his work. Whoever wishes to partake of his flesh will die, for his flesh is made of fornication, greed, desire, and hostile times [cf. Matt. 15:19]. For this reason Herod is likened to a fox [cf. Lk. 13:32]. And the scribe heard the Savior say, "The foxes have holes" [Matt. 8:20]. And in the Song of Songs, "Catch us the little foxes that spoil the vineyards" [S. of S. 2:15]. And David in Psalm 62 said, "They shall be prey for foxes" [Ps. 63:10]. Physiologus, therefore, spoke wisely of the fox.

## XIX. *On the Peridexion Tree and the Doves*

There is a tree found in India called the peridexion tree, the fruit of which is indeed quite sweet and pleasant. The doves delight in the fruit of that tree and live in it while eating its fruit. The dragon, who is the enemy of the doves, fears the tree and its shade in which the doves dwell and is unable to approach the doves or the shade. If the tree's shadow falls to the west, the dragon flees to the east. If, on the other hand, its shadow falls on the east, he flees to the west. If it happens that a dove is found outside the tree or its shadow and the dragon discovers it, he kills it.

The Father of all is the tree and Christ the Son is the shade, just as Gabriel said to Mary, "Do not be afraid, Mary, the Holy Spirit will come upon you, and the power of the Most High will overshadow you" [Lk. 1:30 and 35]. The fruit is heavenly wisdom and the dove is the Holy Spirit. Beware, O man, lest after you have received the Holy Spirit (that is, the intelligible, spiritual dove descending from heaven and remaining over you), you should be outside the divine being, a stranger to the Father, Son, and Holy Spirit, for the dragon (that is, the devil) will kill you. The devil is unable to approach the tree or the shadow or the fruit. And, if you have the

Holy Spirit, the dragon (that is, the devil) will be unable to approach you.

[We Christians know this peridexion tree around which all things are "right" and in which nothing is "left." The "right" is the only-begotten Son of God. As the Lord said, "For the tree is known by its fruit" [Matt. 12:33]. The shade of the tree is the Holy Spirit, as Gabriel said to Holy Mary, "The Holy Spirit will come upon you, and the power of the Most High will over-shadow you" [Lk. 1:35]. The doves are all faithful people, as the Lord says in the Gospel, "Be simple as doves and clever as serpents" [Matt. 10:16]. Be simple so that you deceive no one, and clever lest snares trip you up. Be vigilant, O man of God, remain in the Catholic faith, continue in it, abide in it, dwell in it, and persevere in our faith of the Father, Son, and Holy Spirit and in one Catholic Church as the Psalmist said, "Behold how good and pleasant it is when brothers dwell in unity" [Ps. 133:1]. And elsewhere, "God makes the harmonious dwell in one house" [Ps. 68:6].]

We have spoken wisely, therefore, of the tree and its fruit and shade.

## xx. *On the Elephant*

There is an animal called the elephant whose copulating is free from wicked desire. The tragelaphus, however, is a different animal. If the elephant wishes to produce young, he goes off to the east near paradise where there is a tree called the mandrake. And he goes there with his mate, who first takes a part of the tree and gives it to her husband and cajoles him until he eats it. After the male has eaten, [they join together] and the female immediately conceives in her womb. And when the time comes for her to give birth, she goes to a pond and

where the water reaches her dugs (that is, the female's) she brings forth her offspring (that is, her young). Since the calf is born in the water, it swims about and finds the thighs of its mother and is suckled at her teats. The male elephant guards her while she gives birth because of the serpent who is an enemy to the elephant. [If the female gave birth outside the water, the serpent would steal her calf and devour it. Therefore, she goes into deep water and gives birth there. The male does not leave her but rather guards her while she gives birth because of the serpent who is an enemy to the elephants.] If the elephant finds a serpent, he kills it by trampling on it until it dies.

This is the nature of the elephant: if he should fall, he is unable to get up again. But how can he fall since he rests against a tree? The elephant has no knee joints enabling him to sleep lying down if he wanted to. Shortly before the beast arrives at the tree against which he is accustomed to rest, the hunter who wishes to capture the animal cuts partly through the tree. When the elephant comes and rests against the tree, both tree and beast fall at the same time. The elephant then cries out and immediately there comes a great elephant who is unable to lift the first. Then they both cry out, and twelve other elephants arrive, and not even they can lift the one who has fallen. Then again, they all cry out, and suddenly a tiny elephant appears who puts his trunk under the great one and lifts him up. It is one of the natural qualities of the tiny elephant that, where any part of his hair or bones is burned, neither the dragon nor any other evil thing may come. [Thus, whoever has the works and commands of God within himself purifies his heart, and no thought of the devil can enter. And whatever harmful wickedness was there will immediately vanish leaving no harmful spirit, hostile thought, or misdeed of the devil.]

The great elephant and his wife represent the persons of Adam and Eve. While in a state of virtue (that is, while they pleased the Lord), before their transgression, they had no knowledge of copulation, nor any awareness of the mingling of their flesh. When, however, the woman ate of the tree (that is, the intelligible mandrake) and gave to her husband, she became big with evil; because of this act they were expelled from paradise. Yet as long as they were in paradise Adam had no knowledge of Eve. This is clear since thus it is written, "And after Adam and Eve were expelled from paradise, then Adam knew his wife, and she conceived and brought forth Cain" [Gen. 4:1] over the blameworthy waters. And David said, "Save me, for I have entered the water up to my soul" [Ps. 69:1]. Immediately, the dragon overthrows them and makes them strangers to virtue (that is, by not pleasing God). And they cry out, calling on God and a great elephant comes (that is, the

32

Law) and does not lift them. Indeed, even as the priest did not lift up the one fallen among thieves [cf. Lk. 10:30]. Nor did the twelve elephants (that is, the chorus of prophets) raise him up, even as the Levite failed to raise up the one wounded by thieves [cf. Lk. 10:32]. But the holy intelligible elephant (that is, the Lord Jesus Christ) did so. Although he is greater than all the rest, he was made small in comparison to them. "For he humbled himself and became obedient unto death" [Phil. 2:8], in order to raise up man. He is the intelligible Samaritan who raises us up onto his breast (that is, the body). "For he himself has borne our infirmities and carried our weaknesses" [Is. 53:4]. In Hebrew Samaritan means guardian and David said of him in Psalm 114, "The lord is guardian over the little ones" [Ps. 113:7]. Where my Lord is present, neither the dragon nor any other evils can approach. Physiologus spoke wisely, therefore, of the elephant.

## XXI. *On Amos the Prophet*

Amos the Prophet says, "I was no prophet, nor a prophet's son, but a keeper of goats" [Am. 7:14]. The Savior says of himself, "I was not a prophet, but God the first-born son, the word within God, in the bowels of the Father." And Isaiah says, "Since you are God and in you is God," thus he says, "You are not the son of a prophet, but the son of the living God" [cf. Is. 45:14]. Yet he was a keeper of goats, since being sent from the lap of the Father he assumed human flesh and became a keeper of goats, that is, of all mankind living in sin. Those who received him and believed in the one who sent him became his flock. Whoever did not receive him and remained in sin are goats pasturing in the desert, as are the Jews today. Tearing asunder the body of Christ, they fastened it to the gibbet of the cross. But Christ destroyed all the sins of our flesh and the author of sin, death, and restored us to life as the Apostle says, "For God sending his own son in the likeness of sinful flesh and on account of sin, he condemned sin to death" [Rom. 8:3]. When he was pierced by the lance, blood and water flowed from his side for his people as a bath of regeneration into eternal life.]

## XXII. *On the Roe*

There is an animal in the mountains who is called *dorchon* in Greek and *caprea* in Latin. The beast loves the high mountains but finds her food in the foothills of the mountains. She sees from far off all who approach her, and she knows whether they come with guile or with friendship.

The roe represents the wisdom of God who loves the prophets, that is, the high mountains toward which the Prophet has raised his eyes. "I have lifted my eyes up to the mountains," he said, "whence my help will come" [Ps. 121:1]. Of the roe Solomon said in the Song of Songs, "Behold, my cousin comes leaping over mountains, bounding over the hills" [S. of S. 2:8]. [Thus, the roe feeds in the valleys just as our Lord Jesus Christ feeds in the Church, since the good works of Christians and the gifts of the faithful are the food of Christ. He said, "For I was hungry and you gave me food, I was thirsty and you gave me drink" [Mt. 25:35], and so on. There are valleys and mountains throughout the world. These are to be understood as the churches in various places.]

The roe, however, leaps over the prophets, bounding over the hills (that is, the apostles). She has keen vision signifying that the Savior sees everything that is done.

He is called God because he sees all things, and from afar he sees those approaching him with guile. He knew that Judas was coming to betray him with a kiss [cf. Lk. 22:48]. Furthermore, it is written, "The Lord knows those who are his" [II Tim. 2:19]. And John said, "Behold, the Lamb of God, behold he who takes away the sins of the world" [John 1:29].

### XXIII. *On the Agate-stone*

There is a stone called the agate. Artisans looking for pearls use this stone to find them. Divers tie the agate to a very strong cord and send it down into the sea. When the agate comes to the pearl, it stops and does not move. Thereupon, following the cord, the divers find the pearl.

### XXIV. *On the Oyster-stone and the Pearl*

I will tell you how the pearl is born. There is a stone in the sea called the oyster. It comes out of the sea early in the morning ahead of the light, and, opening its shell (that is, its mouth), it swallows the heavenly dew and the rays of the sun and moon and the light from the stars above. And thus is born the pearl from the most high celestial bodies.

Here, as in the case of the agate, John himself shows us that the intelligible pearl is Jesus Christ our Lord, saying, "Behold, the Lamb of God, behold, he who takes away the sins of the world" [John 1:29]. The sea is the world and the divers who bring up the pearl are the chorus of holy doctors. Because of ill will, however, sinners carry the pearl back down, so greatly does it

oppose them. My Lord, the Savior, however, is found
intelligibly receiving food from the most high places in
the middle of the shell or two wings, that is, between
the Old and the New Testaments. My Lord has said,
"My kingship is not of this world" [John 18:36] but from
the eternal Father and his holy powers. Physiologus,
therefore, spoke wisely of the agate and the pearl.

[The stone which is called the conch is a figure of Holy
Mary concerning whom Isaiah prophesied, "There shall
come forth a shoot (*virga*) from the stump of Jesse" [Is.
11:1]. And again, "Behold a young woman (*virgo*) shall
conceive within her womb and give birth" [Is. 7:14].
And Holy Mary was called a virgin (*virgo*) because of
this shoot (*virga*). In truth, the flower born to Holy Mary
is our Lord God Jesus Christ. For, just as the stone rises
from the sea, so did Holy Mary rise from the house of
her father to the temple of God and there she received
the heavenly dew. The Archangel Gabriel spoke these
words to her, "The Holy Spirit will come upon you, and
the power of the Most High will overshadow you, therefore the child to be born will be called holy, the Son of
God" [Lk. 1:35]. These words are celestial dew, just as
formerly the holy patriarch Jacob blessing his son said
(indicating that Christ would be born of his seed), "May
God give you of the dew of heaven and of the fatness of
the earth" [Gen. 27:28]. By this he meant the chaste and
unblemished Virgin Mary. Now the morning hour of
which he spoke refers to the time for morning prayer.
The opening of the mouth of the conch indicates Mary's
saying to the Angel, "Behold, I am the handmaiden of
the Lord; let it be to me according to your word" [Lk.
1:38]. She immediately received the Holy Spirit within
her, and the power of the Most High like the sun of justice shone upon her. And in him that was born of her is
life and "the light came which enlightens every man
coming into the world" [John 1:19]. And Paul said, "He
is the splendor of glory and the image of his substance"

[Hebr. 1:3]. And elsewhere, "For in him all the fullness of God was pleased to dwell" [Col. 1:19].

Concerning the pearl, therefore, we read in the Gospel, "The kingdom of heaven is like a merchant in search of fine pearls who, on finding one pearl of great value, went and sold all that he had and bought it" [Matt. 13:45–46]. This merchant is assuredly the chorus of apostles, since he calls all the apostles one merchant because of the unity of their faith. For, "There is not Jew and Greek, slave and free man, Scythian and barbarian, male and female for we are all one in Christ Jesus" [Col. 3:11]. Hence the good and wise merchant (the holy chorus of apostles) seeks good pearls, that is, the Law and the Prophets. Or, each soul believing in God seeks the good pearls, namely the apostles and prophets and patriarchs through whom he is able to come upon that true and precious pearl. Those men are holy stones which turn up on earth. When the good man deals in those stones, he will find the pearl, that is, our Lord Jesus Christ, Son of the living God. And, having sold all that he has, he will buy it disregarding and scorning not only the material things of this life but also his wife and sons and all carnal kinship in addition to his body and soul. The truth is, "He who loses his soul for my sake will find it" [Matt. 10:39]. The chorus of apostles seeing all these things accepts neither gold nor silver. As holy Peter said to the lame man asking for alms, "I have no silver and gold, but I give you what I have. In the name of our Lord Jesus Christ arise and walk" [Acts 3:6]. And Paul said, "Everything that was profitable I have deemed a loss on account of Christ and his lofty innocence" [Phil. 3:8]. Who would deliberately despise all his possessions, his wife and sons and all his kin in addition to his body and soul because of the acquisition of a single pearl, unless he firmly confessed and believed that through that one pearl he might acquire greater and better riches, more distinguished honor and a crown of glory? All that the merchant (who is the chorus of apostles) possesses, he has through that one precious stone, the Lord Jesus Christ who is the true pearl, the way, the truth, and our life.

I have heard Jesus say in the Gospel, "Behold, I have given you power over unclean spirits to tread upon serpents and scorpions and every diabolical power and to cure every disease and every infirmity" [Lk. 10:19 and Matt. 10:1]. And again, "And preach as you go since the kingdom of heaven approaches. Heal the sick, cleanse the lepers, cure the blind, raise the dead, cast out demons" [Matt. 10:7–8]. See how inestimable is that pearl to the holy martyrs who performed wonders not only while in this life but also after leaving it, just as we now see how unclean spirits within the possessed are tortured and punished by the virtue and power of the martyrs, and how they are lashed by invisible scourges until they are cast out and put to flight by men. We hear the demons loudly crying out, asking men to stop torturing them. But, since they are all different and many-shaped, some cry out, some roar, and others hiss like serpents and flee from the possessed bodies of men, through the powers of the apostles and all saints which have been given them by the Lord according to their merits. They have attained that eminent honor which transcends all earthly honor through that precious stone, for which they gave up all their possessions so that they might possess that heavenly treasure, like those who said to the Savior, "Lo, we have given up our wives and sons and all our possessions on your account. What will you do for us in your kingdom? Jesus said to them, Truly, I say to you, when the Son of man shall sit on his throne of majesty to judge the world, you will also sit on twelve thrones, judging the twelve tribes of Israel" [Matt. 19:27–28]. For this reason the apostle

Paul confidently states, "Do you know that we shall
judge the angels? And by us the world will be judged"
[I Cor. 6:2].

With such great glory and honor was the chorus of
apostles rewarded that even while still in this world
Paul, that true athlete of Christ, foresaw the crown of
his righteousness in heaven. Rejoice, he said, "I have
fought the good fight. I have finished the race, I have
kept the faith. Henceforth there is laid up for me the
crown of righteousness, which the Lord, the righteous
judge, will award to me on that Day, and not only to me
but also to all who have loved his appearing" [II Tim.
4:8]. The chorus of blessed apostles merits such a crown
from Christ and will receive such a reward in place of
corruptible things.]

## xxv. *On the Adamant-stone*

Adamant-stones have a different nature. Adamant neither
fears iron nor absorbs the smell of smoke. Neither
a demon nor any other evil thing approaches the house
where it is found. It is found in the houses of kings.
Whoever possesses it conquers every man and beast.

My Lord is adamant-rock. If you possess him, no evil
will befall you.

## xxvi. *On the Other Nature of the Wild Ass and the Monkey*

There is another nature to the wild ass, as the wise one
said, since he is found in regal houses, and on the
twenty-fifth day of the month of Famenoth people
know from the wild ass that it is the equinox. Indeed, if

the beast brays twelve times, the king and his palace
know the equinox (*ysemaria* in Greek) is at hand. It is the
equinox also if the monkey makes water seven times.
[Physiologus says of the wild ass that, on the twenty-
fifth day of the month of Famenoth or March, he brays
twelve times in the night and again in the day. By this
is known that it is the equinox of day or night. And they
know the time, hour by hour, by the braying of the ass
once per hour.

The wild ass represents the devil since, when he
knows that night and day are equal (that is, when the
devil perceives that those walking in the shade and
shadow of death have turned lately to the living Lord
and equal the faith of the patriarchs and prophets as
night equals day), then he brays day and night, hour by
hour, seeking the food which he has lost. Now the wild
ass brays only when receiving food, as Job said, "Never
without cause does the wild ass bray but only when
wanting fodder" [cf. Job 6:5]. Similarly, the apostle
Peter says of the devil, "Our adversary prowls around
like a roaring lion, seeking someone to devour" [I Peter
5:8].

The monkey represents the very person of the devil
since he has a beginning but has no end (that is, a tail).
In the beginning, the devil was one of the archangels,
but his end has not been found. [He has no tail since,
just as he perished in the beginning in heaven, so also
will he perish utterly at last, as Paul, the herald of truth,
said, "The Lord Jesus will slay him with the wrath of his
mouth" [II Thes. 2:8].] It is fitting also that, in addition
to not having a tail, the monkey lacks beauty also. And
he is quite ugly in the region where he lacks a tail. Just
so the devil has no good end. Physiologus, therefore,
spoke well.

### XXVII. *On the Indian-stone*

There is a stone called *sindicus* which has the following nature: if a man is dropsical, the doctors seek out this stone, and when they find it, they tie it to the dropsical man for three hours, whereupon the foulness leaves him and is absorbed into the stone itself. When they have untied the stone from the man, they weigh them both and the moderate-sized stone outweighs the man in the scales. If it is exposed to the sun for three hours, however, the stone pours forth the most foul water, which it received from the body of the dropsical man, and becomes clean again just as it was before.

The stone is our Lord Jesus Christ. Because we were dropsical, having the waters of the devil in our hearts, the stone of the Savior came down and was bound and his love surrounded our hearts. Rising from the dead, however, he took away from our soul every intelligible infirmity. "And he himself bore our infirmities" [Matt. 8:17].

### XXVIII. *On the Heron, that is, the Coot*

There is a bird called the heron about which David was aware when he said, "The heron is leader of their house" [Ps. 104:17]. This is the coot of Psalm 104. This beast is prudent beyond all other birds for it does not seek many nests, but, where it settles, there too it feeds and returns there and sleeps. Nor does this bird feed on carrion or fly off to many different places. Her nest and food are in one place.

And you, O citizen, let the Holy Catholic Church be your one and eternal nurse so that spiritual food and heavenly bread become easily digestible within you. Do not seek many places of foreign glory (that is, of heretics).

## XXIX. *On the Fig Tree*

Amos said, "I am not a prophet, nor the son of a prophet, but a goatherd gathering figs" [Am. 7:14], a herder of she-goats and one who feeds the bucks. Amos represents well the person of Christ the Savior. But "gathering figs" is an intelligible and spiritual expression. Zacchaeus climbed up into a fig tree [cf. Lk. 19:4]. But he says "gathering figs" (or scraping them for holy men) since before they have been scraped there are insects living within the figs in darkness, seeing no light and saying among themselves, "We live in a spacious region." But they are sitting in the darkness [cf. Matt. 4:16]. When the fig is broken open, however, with an iron tool and they come out, they see the splendor of the blazing sun and the moon and the stars and say, "We were in the darkness, sitting in the shadow" [cf. Is. 9:2 and Matt. 4:16]. Isaiah said of these, "Let him bring out the chained from their chains, and from the prison those who live in the darkness" [Is. 42:7] Gather them on the first day, and on the third day the fruit will be mature and there will be food for men.

My Lord's side was pierced, that is, opened with the lance and the sword, and blood and water came forth. On the third day, rising up from the dead, we perceive the intelligible and spiritual lights. In just such a fashion the midges (that is, the insects) see the immortal lights when the fig is opened. The goats represent the person doing penance, since their hair is thrown into a sack and "Among them they do their penance in ashes" [Matt. 11:21 and Lk. 10:13]. "The people who sat in darkness have seen a great light, and for those who were in the shadow of death light has dawned" [Matt. 4:16]. On the third day after the fig has been opened there will be food. Our Lord was struck down and on the third day, rising up from the dead, food and life were made for us all.

## xxx. *On the Panther*

The Prophet says, "I became like a lion to the house of
Judah, and like a panther to the house of Ephraim"
[Hos. 5:14]. The panther has this nature: he is a friend
of all animals but is an enemy of the dragon. He is entirely variegated and is beautiful like Joseph's cloak [cf.
Gen. 37:3]. And David said in Psalm 44, "At your right
hand stands the queen in a golden dress decked in various hues" [Ps. 45:13–14]. The panther is a quiet and
most exceedingly mild animal. If, however, he has eaten
and is satisfied, he falls asleep immediately in his lair
and arises from his sleep on the third day (like our
Savior). If the panther awakens from his sleep on the
third day, he roars out in a loud voice and many a pleasant fragrance issues from his voice. Those who are far
away and those who are near, hearing his voice, follow
its pleasant fragrance.

Our Lord and Savior rising up from the dead became
a pleasant fragrance for us [cf. II Cor. 2:15], peace for
those who are far away and those who are near [cf.
Eph. 2:17]. As the apostle Paul said, "Manifold is the
wisdom of God [cf. Eph. 3:10], for it includes virginity,
abstinence, mercy, faith, love, harmony, peace, joy, and
patience" [cf. Gal. 5:22]. The heavenly wisdom of
Christ who is God is variegated with each form of love.
It is wisely said that the panther is an enemy of the
dragon in the water. Therefore, Holy Scriptures have
said nothing concerning birds and animals without the
purpose of our understanding. As the Apostle said
about Satan, "We are not ignorant of his cunning" [II
Cor. 2:11], for he travels every path which is not good
[cf. Jer. 6:16].

[Only the dragon when he hears the panther's voice
is seized by fear and bolsters itself within subterranean
caves where it does not suffer the power of that sweet
fragrance. Having coiled itself up, the dragon falls into a

deep sleep and remains there immobile and senseless as if dead. The other animals, however, follow the panther wherever he goes.

Thus, our Lord Jesus Christ who is the true panther draws to himself all humankind (captured by the devil and held punished by death) through his incarnation. "He led a host of captives" [Eph. 4:8]. As David the Prophet said, "Thou didst ascend the high mount, leading captives in thy train, and receiving gifts among men" [Ps. 68:18]. Now the word panther means "gathering all things," just as our Lord God, as we have said, seeing humankind captured by demons and given over to idols, and all nations and peoples made prey to the devil, coming down from heaven snatched us from the power of the devil and joined us to his goodness. He carried away the sons of filial piety and fulfilled what the Prophet predicted, "I have become like a panther to Ephraim, and like a lion to the house of Judah" [Hos. 5:14] which served idols. By this is meant the call to the Gentiles and Jews.

The panther is a variable beast as Solomon said of the Lord Jesus Christ, "who is the wisdom of God, the intelligible spirit, holy, one, manifold, subtle, mobile, certain, uncontaminated, true, sweet, loving the good, proper, preventing no good thing from coming about, mild, firm, subtle, secure, all-powerful, all-seeing, all-doing, more agile than wisdom," and so on [Wisdom 7:22]. Paul, the doctor of truth, witnesses that Christ is divine wisdom: "But we preach Christ crucified, a stumbling block to Jews and folly to gentiles, but to those who are called, both Jews and gentiles, Christ is the power of God and the wisdom of God" [I Cor. 1:23–24].

And since the panther is a beautiful animal David said of Christ, "You are the fairest of the sons of men" [Ps. 45:2]. And because it is a very mild animal Isaiah said, "Rejoice and be glad, daughter of Zion. Tell, O daughter of Jerusalem, that your king comes to you mild and saving" [cf. Is. 62:11]. Just as the panther, having eaten and been filled, immediately lay down and slept, so also our Lord Jesus Christ, after he was satiated by Jewish illusions, that is, by whips, blows, injuries, reproaches, thorns, and spit, was hung on the cross by his hands, pierced with nails, his thirst quenched with vinegar and gall, and in addition his side was pierced with a spear. Satiated with all these great Jewish rewards, Christ slept and rested in the tomb and descended into hell and there bound the great dragon, our enemy.

And as that animal rose up on the third day from sleep and sent forth a great roar with sweet fragrance flaming from his mouth, so also did our Lord Jesus Christ arise from the dead on the third day as the Psalmist said, "Then the lord awoke as from sleep, like a strong man shouting because of wine" [Ps. 78:65]. Immediately, he cried out with a loud voice so that the sound was heard in every land as his words at the ends of the world: "Be of good cheer and do not fear, for I have overcome the world" [John 16:33]. Again, "Holy Father, those whom thou hast given me I have guarded, and none of them is lost but the son of perdition" [John 17:12]. And, "I go to my Father and your Father, to my God and to your God" [John 20:17]. And again, "I will come to you, and I will not send you away orphans" [John 14:18]. And at the end of the Gospel he says, "Lo, I am with you always, to the close of the age" [Matt. 28:20].

Thus, as the sweet fragrance goes forth from the mouth of the panther, both those close and those far away (i.e., the Jews, who sometimes having the sense of wild beasts were near to him through the Law, and the Gentiles, who were far away without the Law) hearing his voice were filled and renewed with the very sweet fragrance of his commands. They followed him, calling out with the Prophet and saying, "How sweet are thy

words to my taste, lord, sweeter than honey and the
honeycomb to my mouth" [Ps. 119:103]. David said of
the fragrance of his commands, "Grace was poured
upon your lips; therefore, God has blessed you forever"
[Ps. 45:2]. And Solomon said of it in the Song of Songs,
"The fragrance of your oils is more than all other spices"
[S. of S. 4:10]. What else can the oil of Christ be but his
commands spiced beyond all others? For just as a strong
kind of spice gives off a sweet fragrance, so too the
words which issue from the mouth of the Lord rejoice
the hearts of men who hear and follow him. "Your
name is oil poured out; therefore the maidens love you"
[S. of S. 1:3]. "They draw you after them; we run in the
fragrance of your oils" [S. of S. 1:4]. And shortly thereafter, "The king has brought me into his chamber" [S. of
S. 1:4]. Let us run as quickly as the maidens (i.e., those
souls renewed in baptism) after the oils of Christ's commands. Let us pass from earthly to heavenly regions so
that the king might lead us into his palace, that is, into
Jerusalem, the city of God, and into the mountain of all
saints. And when we have earned entrance, let us say,
"Glorious things are spoken about you, O city of God"
[Ps. 87:3]. "As we have heard, so have we seen in the
city of the Lord of powers" [Ps. 48:8]. Physiologus
speaks well of the panther.]


## xxxi. *On the Whale, that is, the Aspidoceleon*

Physiologus spoke of a certain whale in the sea called the
aspidoceleon that is exceedingly large like an island,
heavier than sand, and is a figure of the devil. Ignorant
sailors tie their ships to the beast as to an island and
plant their anchors and stakes in it. They light their
cooking fires on the whale but, when he feels the heat,
he urinates and plunges into the depths, sinking all the

ships. You also, O man, if you fix and bind yourself to the hope of the devil, he will plunge you along with himself into hell-fire.

The whale has another nature: when he grows hungry he opens his mouth very wide and many a good fragrance comes out of his mouth. Tiny little fish, catching the scent, follow it and gather together in the mouth of that huge whale, who closes his mouth when it is full and swallows all those tiny little fish, by which is meant those small in faith. We do not find the larger and perfect fish approaching the whale, for the perfect ones have achieved the highest degree. Indeed, Paul said, "We are not ignorant of his cunning" [II Cor. 2:11]. Job is a most perfect fish as are Moses and the other prophets. Joseph fled the huge whale, that is, the wife of the prince of the cooks, as is written in Genesis [Gen. 39]. Likewise, Thecla fled Thamyridus, Susanna the two wicked old men of Babylon, Esther and Judith fled Artaxerxes and Holofernes. The three boys fled the King Nebuchadnezzar, the huge whale [Dan. 3], and Sara the daughter of Raguelis fled Nasmodeus (as in Tobia). Physiologus, therefore, spoke well of the aspidoceleon, the great whale.

## xxxii. *On the Partridge*

Jeremiah said of the partridge, "The partridge cried out, gathering the brood which she did not hatch" [cf. Jer. 17:11]. The partridge warms the eggs of other birds, laboring over them and nourishing them. When the chicks have grown up and begin to fly, however, each kind flies away to its own parents and leaves the partridge by herself.

Thus the devil seizes the stock of little ones but, when they reach the fullness of age, they come to Christ and

the church, and the devil proves a fool. [When the young hear the voice of Christ, they use their wings made spiritual through faith to fly away and commend themselves to Christ. He immediately receives them with great fatherly reward and love in the shade of his wings and gives them to the church to be nourished.] If today one has evil ways, tomorrow he will become sober. Flee the devil, that is, the partridge, and you will come to your rightful parents, the prophets and apostles.

## XXXIII. *On the Vulture*

Physiologus says of the vulture that she will be found in lofty and high places and sleeps on lofty rocks and on the pinnacles of temples. When she becomes pregnant, however, the vulture goes to India and gets the *eutocius* stone. This stone is similar to a nut in size and, if you shake it, another stone within it will move and sound like a bell. And, when the time for bearing comes upon the vulture, she sits on this stone (called the *eutocius*) and gives birth without pain.

And you, O man, if your soul is made big with the wicked sorrows of the devil who is our adversary, take within you the intelligible stone *eutocius*. Its exterior is God-bearing. Mary internally bore the spiritual stone, our Savior. "The stone which the builders rejected has become the cornerstone" [Ps. 118:22]. And, "A stone cut out of the mountain without hands" [Dan. 2:34] was born and rolled up into good hope because of our old errors. And also he was crucified for the redemption of our sins. Then all acts of adultery, fornication, drunkenness, and others are cleansed from your soul and then [you will receive] the word of the heavenly King. This is what Isaiah said: "In fear of you, O Lord, we conceived and brought forth the spirit of your deliverance which you accomplished on earth" [cf. Is. 26: 17–18]. Indeed, even the Old Testament had in its ministry a Savior, yet the Savior was hidden from the false Jews and appeared, therefore, to us. Therefore, this was spoken well of the vulture and the stone.

With this stone David slew Goliath. Aaron and Hur supported Moses' hands with two stones when he put Amaleck to flight [Ex. 17:12]. The vulture does not have a single dwelling or nest, and we do not have the nest of old idolatry (that is, we used to pursue the worship of idols and of many gods) but faith in the church from which heavenly paternal grace has appeared. Thus, we have been saved by Jesus Christ.


## xxxiv. *On the Ant-lion*

In Job, Eliphaz King of the Temanites says of the antlion, "He perished because he had no food" [Job 4:11]. His father has the face of a lion and eats flesh, while his mother has the face of an ant and feeds on plants. If she brings forth an ant-lion, it perishes because it has two natures, the face of a lion and the fore and rear parts of an ant. Because of the mother's nature, it cannot feed on flesh nor can it eat plants because of the father's nature. It perishes, therefore, because it has no food.

So it is with each person: "The man of deceitful heart is confused in all his ways" [Jas. 1:8]. It is not proper, therefore, to follow two paths, O man of double mind, even in prayer to be a sinner following two paths. It is written wisely, "Let it be with you *yes* or *no*" [Matt. 5:37].

## xxxv. *On the Weasel*

The law says, "You shall not eat the weasel, nor anything like it" [Lev. 11:29]. The weasel has this nature:
the female receives the seed of the male in her mouth
and, having become pregnant, gives birth through her
ears. If it happens, however, that she gives birth
through the right ear, the young will be male, and if
through the left ear, female. Wicked things are en-
gendered through the ears.

There are those even now eating the spiritual bread in
church. When they have been dismissed, however, they
will cast the Word out of their hearing. They will become as Psalm 57 says, "Like the deaf viper who stops
his ears, and will not hear the voice of the enchanter and
wizard, and be charmed by the wise one" [Ps. 58:4–5].

[Physiologus says that the viper has the following nature: when a man approaches the cave where the viper
dwells and charms it with songs to force it out of its
cave, the beast lays its head on the ground, pressing one
ear against the earth and stopping the other with its tail
so as not to hear the voice of the enchanter.

Such are the rich of the world who lay one ear to
earthly desires and stop up the other, adding new sins
to past ones. Thus, they do not hear the voice of the enchanter (that is, the preachers). Indeed, they blind their
eyes with earthly desires and rapine, so that they desire
neither to hear nor to serve divine commands with their
ears nor to regard heaven with their eyes nor even to
consider him who is above the heavens and perform
works of goodness and justice. Those who do not wish
to hear God now through the preachers and Holy Scriptures will hear him on the Day of Judgment saying,
"Depart from me, you cursed, into the eternal fire prepared for the devil and his angels" [Matt. 25:41].]

### XXXVI. *On the Unicorn*

In Deuteronomy Moses said while blessing Joseph, "His beauty is that of the firstling bull, and his horns are the horns of the unicorn" [Deut. 33:17]. The monoceras, that is, the unicorn, has this nature: he is a small animal like the kid, is exceedingly shrewd, and has one horn in the middle of his head. The hunter cannot approach him because he is extremely strong. How then do they hunt the beast? Hunters place a chaste virgin before him. He bounds forth into her lap and she warms and nourishes the animal and takes him into the palace of kings.

The unicorn has one horn because the Savior said, "I and the Father are one" [John 10:30]. "For he has raised up a horn of salvation for us in the house of his servant David" [Lk. 1:69]. Coming down from heaven, he came into the womb of the Virgin Mary. "He was loved like the son of the unicorns" [cf. Ps. 22:21] as David said in the psalm.

[He is said to be shrewd since neither principalities, powers, thrones, nor dominations can comprehend him, nor can hell hold him. He is small because of the humility of his incarnation. He said, "Learn from me; for I am gentle and lowly in heart" [Matt. 11:29]. He is so shrewd that that most clever devil cannot comprehend him or find him out, but through the will of the Father alone he came down into the womb of the Virgin Mary for our salvation. "And the Word became flesh and dwelt among us" [John 1:14]. The unicorn is like the kid, as is our Savior according to the Apostle: "He was made in the likeness of sinful flesh and for sin he condemned sin in the flesh" [cf. Rom. 8:3]. This was spoken well of the unicorn.]

## XXXVII. *On the Beaver*

There is an animal called the beaver who is extremely inoffensive and quiet. His genitals are helpful as a medicine and he is found in the king's palace. When the beaver sees the hunter hastening to overtake him in the mountains, he bites off his own genitals and throws them before the hunter. If another hunter happens to pursue him later on, he throws himself on his back and shows himself to the hunter. And the hunter, seeing that the beast has no genitals, departs from him.

O, and you who behave in a manly way, O citizen of God, if you have given to the hunter the things which are his, he no longer approaches you. If you have had evil inclinations toward sin, greed, adultery, theft, cut them away from you and give them to the devil. The Apostle said, "Pay all of them their dues, taxes to whom taxes are due, honor to whom honor is due," and so on [Rom. 13:7]. Let us first throw the disgraces of sins which are within us before the devil, for they are his works, and let us give to God the things which are God's, prayers and the fruit of our good works. [Separate yourself from the works of the flesh, which are the tax and tribute of the devil, and acquire the spiritual fruits: charity, joy, peace, patience, goodness, faith, meekness, continence, chastity in good works, that is, in alms, visitations to the sick, the care of the poor, the praise of God, in prayers and the performance of kindnesses and other acts of God.]

## XXXVIII. *On the Hyena or the Brute*

There is an animal which is called the hyena in Greek and the brute in Latin.] The Law said, "Thou shalt not eat the brute, nor anything similar to it" [cf. Lev. 11:27].

This animal is an *arenotelicon*, that is, an alternating male-female. At one time it becomes a male, at another a female, and it is unclean because it has two natures. Therefore, Jeremiah said, "Never will my heritage be to me like the cave of the brute" [cf. Jer. 12:9].

Thus double-minded men are compared to the brute. They have the nature of men, that is, courage at the signal for gathering the congregation together, but when the assembly is dismissed they take on womanly nature. [The sons of Israel are like this animal since in the beginning they served the living God but later, given over to pleasure and lust, they adored idols. For this reason, the Prophet likens the synagogue to an unclean animal. Whoever among us is eager for pleasure and greed is compared to this unclean brute since he is neither man nor woman, that is, neither faithful nor unfaithful. The Apostle said, "The root of all evils is enslavement to idols" [cf. Tim. 6:10]. Solomon said of these without doubt, "A double-minded man is unstable in all his ways" [Jas. 1:8]. The Savior said to them in the Gospel, "You are not able to serve two masters, God and mammon" [Matt. 6:24].] Physiologus spoke well.

## xxxix. *On the Niluus*

There is an animal called the niluus who lives in the river and has the shape of a dog. He is an enemy of the crocodile, and, if he sees the crocodile sleeping with his mouth open, the niluus goes and besmears himself all over with mud. When the mud has dried, he springs into the crocodile's mouth. [Tearing at the crocodile's intestines, he emerges alive from the dead beast after having lacerated his entrails] and expelled his intestines and guts.

Thus is hell to be understood which steals and kills

each soul. Our heavenly Savior, however, accepting an
earthly body, descended into hell until he had stolen
away those who were formerly dead according to the
holy promises, and also to take away the strength and
sting of death. [Christ descended into hell and, bursting
its entrails, led away all who were devoured and de-
tained there. Thus the Evangelist says, "The tombs also
were opened, and many bodies of the saints were
raised" [Matt. 27:52]. Therefore, he put death to death
and, arising living from the dead, he exulted over death
with the words of the Prophet, "O death, I shall be thy
death; and thy bite I shall be, O hell" [Hos. 13:14]. And
elsewhere, "Death is swallowed up in the victory of
Christ; O death, where is thy grief, O death, where is
thy sting?" [I Cor. 15:54–55].]

## XL. *On the Echinemon*

There is an animal called the echinemon which is hostile
to the dragon. If he encounters a dragon, he goes
against him and besmears himself with mud and covers
his nostrils with his tail, hiding himself and inflating
himself, and thus stands against the dragon until he
kills it.

Thus also did our Savior, taking on from the earth the
substance of a body, that is, the body which he received
from Mary, stand until he slew the intelligible dragon
Pharaoh (that is, the devil), who sits by the rivers of
Egypt [Ez. 29:3].

## XLI. *On the Little Crow*

Jeremiah bore witness saying, "You sat like a little crow
who has been abandoned" [cf. Jer. 3:2]. Physiologus

stated that the little crow belongs to but one husband
and, if he dies, she will not become another's nor will
she be joined with another female's husband.

And thus, therefore, the synagogue of the Jews in the
earthly Jerusalem is now an abandoned little crow who
killed its heavenly man in not receiving Jesus Christ and
for this was abandoned. The Apostle said of those
among the Gentiles who believed, "I have decided to
present you to Christ as a pure bride to her one husband" [II Cor. 11:2]. Therefore, we have the divine man
in the Word which is in our minds, so that a stranger
cannot approach us and we be found in the cave of
thieves.

## XLII. *On the Ostrich*

There is an animal called the ostrich which the Greeks call
the *struthiocamelus* and the Latins the *struthio*. Jeremiah
the Prophet says of this animal, "Even the ostrich in the
heavens knows her own time" [Jer. 8:7]. Physiologus
says that this bird is like a vulture. Although she has
wings, she does not fly like other birds; she has feet
like those of a camel, and hence the Greeks call her the
sparrow-camel. When her time comes for laying eggs,
this animal raises her eyes to the heavens to see if that
star called Virgilia is rising, for she does not deposit her
eggs in the ground until that star rises. Job says of that
star, "[He is wise in heart and mighty in strength] who
made Virgilia, the north and the right and the store-
houses of the south" [Job 9:9]. Now in its own time
Virgilia arises in summer around the month of June
when the crops flourish. Then, when the ostrich sees
that Virgilia has risen in the heavens, she digs a hole in
the desert ground, deposits her eggs, and covers them
with sand. Once having departed that place, however,

she immediately forgets it and does not return to her
eggs, for this animal is by nature forgetful. Hence, she
produces eggs in summer and covers them with sand,
which hatches the chicks as the mother might do sitting
on her eggs. The tranquility of the season and the tem-
perateness of the air seem to be more perfectly suited
than the mother for the task, so that when the summer
is hot the sand incubates the eggs and hatches the
chicks.

Thus, if the ostrich knows her own time and raises
her eyes to heaven and forgets her young, how much
more fitting is it for us to know our time, to forget earthly things and pursue heavenly things and to raise the
eyes of our hearts? The Apostle says, "Forgetting what
lies behind, I press on toward the goal for the prize of
the upward call" [Phil. 3:13–14]. And the Lord says in
the Gospel, "He who loves father and mother or children more than me is not worthy of me" [Matt. 10:37].
And to the one who excused himself on account of the
burial of his father he said, "Follow me, and leave the
dead to bury their own dead" [Matt. 8:22].]

XLIII. *On the Turtle-dove*

Solomon said, "The voice of the turtle-dove is heard in
our land" [S. of S. 2:12]. The turtle resides in the wilderness, that is, she withdraws into solitary places for she
does not like to be among many others.

And it is said that our Savior stayed on the Mount of
Olives. Taking with him Peter, James, and John, he
went up the mountain [cf. Mk. 3:13]. Just as the turtle
likes to withdraw into silence, so do generous Christ-
bearers choose to live in secret. Christ-bearers are those
who have put on Christ and who imitate our Lord Jesus
Christ, the turtle-dove.

[Physiologus said of the turtle-dove that she loves her
mate very much and lives chastely and faithfully with
him, such that, if the male happens to be captured by a
hawk or fowler, she does not take another mate but,
rather, longs for and awaits her lost one at every moment and endures thus in remembrance and longing for
him until death.

Take note, therefore, all you souls of the faithful, how
much chastity is found in a small bird. All you who bear
the person of the turtle-dove in the visage of the soul,
imitate her chastity. For such is the holy church which,
after seeing her mate crucified and resurrected on the
third day and ascended into heaven, does not take another mate but longs for him and awaits him enduring
in love and charity until death. Our Lord Jesus Christ
said, "He who endures to the end will be saved" [Matt.
10:22]. Similarly, even the prophet David exhorts us in
the psalm, "Be strong and let your heart take courage,
and yea, wait for the Lord!" [Ps. 27:14].]

## XLIV. *On the Swallow*

Hezekiah the King said in the book of the prophet Isaiah, "Like a swallow I cry out, and like a dove I moan" [Is. 38:14]. And in Jeremiah it is said, "The turtle-dove and the swallow and the stork keep the time of their entry" [Jer. 8:7]. Physiologus said of the swallow that it gives birth only once and never again.

And my Savior was conceived but once in the womb. He was carried once, crucified once, and once rose from the dead: "One God, one faith, one baptism" [Eph. 4:5].

## XLV. *On the Stag*

It is said in Psalm 41, "As the stag longs for flowing streams, so longs my soul for thee, O God" [Ps. 42:1]. The stag is an enemy of the dragon. Moreover, the dragon flees from the stag into the cracks in the earth, and the stag, going and drinking from a stream until his muzzle is full, then spits out the water into the cracks and draws the dragon out and stamps on him and kills him.

Thus did our Lord kill the huge dragon, the devil, with heavenly waters of indescribable wisdom. The dragon cannot bear water, and the devil cannot bear heavenly words. If you also have intelligible dragons hidden in your heart, call upon Christ in the Gospels with prayers and he will kill the dragon. "You are God's temple, and God's spirit dwells in you" [I Cor. 3:6]. You will never find a dragon in the house where the stag's hair appears or where his bones are burned. Likewise, if the traces of God and fear of Christ are found in your heart, no impure spirit will enter you.

[When the devil saw Christ in the country of the Garasenes, he ran by with an army of demons inhabit-

ing the body of one man and said, "What have you to
do with me, O Son of God? Have you come to torment
us before the time?" [Matt. 8:29 and Mk. 5:7]. "The
Lord asked him, What is your name? He replied, My
name is legion" [Mk. 5:9]. "And he begged Jesus not to
order them to go into the abyss. There was a great herd
of swine feeding there. The demons begged him, If you
cast us out, send us into the swine. And Jesus said to
them, Go, and they came out of the men and went into
the swine. Behold, the whole herd went away in great
haste over the steep bank into the sea, almost two thou-
sand of them, and perished in the waters" [Mt. 8:31–32

and Mk. 5:10–13]. Hearing the voice of the Lord, the
devil fled over the steep bank with all his followers. In
more recent days, the Apostle witnessed concerning the
devil and said, "The Lord Jesus will slay him with the
breath of his mouth" [II Thes. 2:8].

And David said, "The high mountains are for the stags" [Ps. 104:18]. He calls the apostles and prophets mountains, and stags he calls the faithful men who attain to knowledge of Christ through the apostles, prophets, and priests. It is written in the psalm, "I have lifted my eyes up to the mountains whence my help will come" [Ps. 121:1].]

## XLVI. *On the Frog*

There is a frog called the *cerseus*, meaning "the one from the dry place." This frog is not bothered by the heat during summer but, if he is caught in the rain, he will die. If the water frogs, however, who live in bodies of water look upon the rays of the sun and become warm, they baptize themselves in a stream.

"The ones from the dry place" represent fine, abstinent men who are unaffected by working patiently in abstinence; however, if they are caught in the rain (that is, in worldly desires), they die. The water frogs, however, are those who cannot stand abstinence. If these abstain until daytime, not being able to bear a ray of intelligible sunlight, they slip back again into their former desires.

### XLVII. *On the Lizard, that is, the Salamander*

The story of the three brothers in Daniel is a wondrous one. They were thrown into the fiery furnace in order that they might sing a hymn to God [cf. Dan. 3]. Yet, this is not such a wonder because they were just men, nor would it be such a great thing if they had raised the dead or moved mountains into the sea [cf. I Cor. 13:2]. Physiologus said of the lizard which is called the salamander that, if it is put into a fiery furnace or an oven for the baths, the fire will be quenched. Such is the salamander's nature. How much better are those "who through justice quenched the power of fire, and who stopped the mouths of lions" [Heb. 11:33].

[Thus everyone who believes in God with all his faith, and endures in good works, will pass through hell-fire and the flame will not touch him. Isaiah the Prophet wrote of this, "If you walk through fire, the flame shall not consume you" [Is. 43:2].]

### XLVIII. *On the Magnet*

There is a stone which is called the magnet which suspends iron if it is close to (that is, if it is put next to) the iron. If a created thing does this, how much better is the creator and maker of all things who suspends heaven from earth and who stretches out heaven like a skin.

62

## XLIX.  *On the Adamant-stone*

T he adamant-stone is found in the regions of the East,
where it does not shine during the day but is found at
night. It is called adamant because it "dominates" all
things, but no one can "dominate" it.

My Lord and Savior judges all men and is judged by
no one. The Prophet says in Amos, "I saw an adamant-
stone" [cf. Am. 7:7]. Thus will you be if you are found
in the regions of the East, and especially if you are called
just, blameless, and pious like Job [cf. Job 1:1], who was
innocent at sunrise, and so forth.

[The Prophet said of this adamantine stone, "I saw a
man standing on a wall of adamant, and in his hand an
adamant in the midst of the people of Israel" [Amos
7:7]. Since creatures cannot prevail against the Creator,
Christ is the adamant. He stands on the wall of adamant, on the holy and living stones of which the heavenly Jerusalem is built. These are the apostles, prophets,
and martyrs against whom neither fire, sword, nor the
teeth of beasts could prevail. Just as the Christians are
named after Christ, so too are all holy men called adamant by the Prophet after that stone. He says, "I saw a
man standing on a wall of adamant, and behold" he
said, "an adamant in his hand," that is, the Son of God
and Son of man who saw fit to assume flesh in the
womb of Mary. He held the stone in his hand as the
glory of his godhead, as Daniel bears witness concerning him saying, "I looked and behold," he said, "a man
clothed in baldachin" [Dan. 10:5]. Now the man signifies the majesty of godhead, and linen the carnal man
whom he saw fit to put on, since baldachin means linen,
that is, a clothing derived from the earth. The Apostle,
blessed Peter, said of his being called a man, "Jesus of
Nazareth a man-God has been made known to us"
[Acts 2:22]. Moreover, blessed Paul said, "I betrothed
you to Christ to present you as a pure bride to her one

husband" [II Cor. 11:2]. That we might know that it is
Christ of whom he is speaking, Paul says, "Do you seek
a proof of him, Christ who speaks within me" [II Cor.
13:3]?

Therefore, the eastern mountain mentioned by Physiologus where the adamant is found signifies God the
unbegotten Father from whom all things come. He calls
the mountain high and his glory inaccessible just as the
apostle Paul speaks about him who alone has immortali-
ty and inhabits the inaccessible light where the stone is
found. "Christ is in the Father and the Father in me"
[John 14:10]. Again, "Who sees me, also sees the Father" [John 14:9].

The fact that the stone is not found during the day
signifies that Christ concealed his descent from the
heavenly virtues and dominations and powers who like
heavenly bodies stand by God. They did not know the
one bearing the just, heavenly office of his descent and
incarnation which was to be on earth. Finally, having
accomplished all his marvels for the redemption of mankind, when he ascended into heaven the multitude of
the lofty city, seeing him a whole and perfect man, said,
"Who is this king of glory who comes up from Edom
in a crimsoned garment from Bozrah?" [Is. 63:1]. Who is
this one who arises out of blood and the red of his clothing from flesh? The stone is found at night since he
came down into the darkness of this world and illu-
mined this entire race which sits in darkness and in the
place of the shadow of death, as the prophet David said,
representing the entire human race, "Yea, Lord, thou
dost light my lamp; my God lightens my darkness"
[Ps. 18:28]. Hence our Lord came down and, taking on
to himself the lamp which the devil extinguished (that
is, the soul and body), he lighted it with the splendor of
his glory, giving it life and openly taking it back with
him. The Apostle said of this sacrament of wondrous
mystery, "Plainly," he said, "great is the mystery of

piety which was manifested in the flesh, vindicated in
the Spirit, which appeared to the angels, was preached
among the nations, believed on in this world, taken up
in glory" [cf. I Tim. 3:16].

Furthermore, Physiologus says of the stone that iron
does not prevail against it, that is, death does not rule it.
He destroyed death and trampled it under foot as the
Apostle bore witness, saying, "Death is swallowed up
in victory. Where is thy struggle, Death? Where is thy
sting?" [I Cor. 15:55]. Nor can fire do anything to this
rock, meaning the devil who with his fiery darts in-
flames every land and the lustful, drunken, wrathful
cities. Of them Isaiah says, "The land lies desolate, your
cities are burned with fire" [Is. 1:7]. "The Lord Jesus
Christ killed him with the breath of his mouth" [II Thes.
2:8]. No other stone harms it, that is, no man at all nor
any creature can stand against him. "All things were
made through him, and without him nothing was
made" [John 1:3].]


L. *On the Doves*

John the Evangelist says, "I saw heaven open, and the
Spirit of God descending on him" [John 1:32]. Physiologus spoke of many kinds of doves [colored, speckled,
black, white, red (*stephanitus*), yellow-gold, sky-blue,
ashen, golden, and honey. The red dove is first among
them and rules over and pacifies them and gathers wild
doves into its dovecot. He too is first who redeems us
through his precious blood, and gathered us from different peoples into the one house of the church. Neither
Moses nor Elias nor any of the prophets or patriarchs
but he alone coming from the Father saved us, re-
deemed us from perpetual death through his passion.

John said, "I saw heaven open and the Spirit of God descending like a dove" [Matt. 3:16] sent by God. "He desires all men to be saved and to come to the knowledge of the truth" [I Tim. 2:4]. Wishing to gather mankind into the holy Catholic and Apostolic Church, he sent the Holy Spirit "speaking in many and various ways" [Heb. 1:1] through the Law and the Prophets to all mankind just as the doves' colors are various.

Black signifies the Law because of its obscure words and arguable meaning. Speckled coloration signifies the diversity of the twelve prophets. Sky-blue signifies Elias who was taken up in a chariot through the sky to heaven. Ashen signifies the prophet Jonah preaching to the Ninivites and performing penance in a hair shirt and ashes. For this the Lord granted him protection over his life. Gold signifies none other than those three boys who, having the true spirit of God, said to King Nebuchadnezzar, "Be it known to you, O king, that we will not serve your God or worship the golden image which you have set up" [Dan. 3:18]. Elisha is the color honey, for he received the mantle from his master Elijah as he went into heaven and deserved to be honored with a double share of his spirit [cf. II Kings 2:13–14].

White represents blessed John the precursor of Christ who had the purity of holy baptism. The prophet Isaiah said of it, "Wash yourselves; make yourselves clean, remove the evil of your thoughts from before my eyes; learn to do good. And if your sins were like scarlet, they shall become white as snow" [Is. 1:16–18]. The Lord said of John, "Truly, I say to you, among those born of woman there was no one greater than John the Baptist" [Matt. 11:11]. The Law and the Prophets preached until the time of John, who pointed out Christ with his finger [cf. Lk. 16:16]. "Behold, the Lamb of God, who takes away the sins of the world" [John 1:29].

Red (*stephanitus*) is Stephen the first martyr who after

accepting the holy spirit merited the vision of Christ at
the right hand of the Father. Red signifies the passion of
the Lord on account of which Raab the harlot put aside
the scarlet sign and was saved in Jerico [Jos. 2:18]. The
Song of Songs says, "Your lips are like the red color of
braid" [S. of S. 4:3]. The Gospel says, "The Jews arrayed the Lord in a scarlet robe" [John 19:12]. And
Isaiah says, "Who is this that comes up from Edom in
crimsoned garments from Bozrah?" [Is. 63:1]. And a
scarlet thread was tied to the hand of Zerah the son of
Judah by the midwife while he was still within his
mother's womb [cf. Gen. 38:28–30]. The Song of Songs
says, "My cousin is white and red" [S. of S. 5:10]. White
signifies virginity and red martyrdom through which all
of us who believe have been saved in the name of the
Father and Son and Holy Spirit who is blessed forever,
Amen].

Thus the blood of my Savior has led everyone to eternal life. Neither Moses because he gave the Law, nor
Isaiah, nor any of the prophets, but the Lord Jesus
Christ, the son of God, coming down saved us with his
holy blood. Because of the scarlet sign his soul was
saved from fornication; and Mary, having been chosen,
accepted the scarlet and purple [cf. Rev. 17:4]. It was
said in the Song of Songs, "Your lips are like a scarlet
cloak, and your speech is lovely" [S. of S. 4:3]. And in
Matthew, "They arrayed him in a scarlet robe" [Matt.
27:28].

LI. *On the Sun-lizard, that is, the Sun-eel*

There is a beast called the sun-lizard, that is, the sun-eel.
When this animal grows old, he is hampered by [weak-
ening of] his two eyes. No longer being able to perceive
the sunlight, he goes blind. What does he do? Moved by

his good nature, he finds a wall facing east, enters a crack in that wall, and gazes eastward. His eyes are then opened by the eastern sun and made new again.

And you, O man, if you have the clothing of the old man [cf. Col. 3:9 and Eph. 4:22], see that, when the eyes of your heart are clouded, you seek out the intelligible eastern sun who is Jesus Christ and whose name is "the east" [cf. Zech. 3:8 and 6:12; Lk. 1:78] in Jeremiah. As the Apostle says, "He is the sun of justice" [cf. Mal. 4:2]. He will open for you the intelligible eyes of your heart, and for you the old clothing will become new.

References found in the notes to classical analogues are taken from the editions in the Loeb Classical Library. In addition, the following editions and translations have been used:

Hermes Trismegistus. *Die Kyraniden*. Edited by Dimitris Kaimakis. Beiträge zur klassischen Philologie, Heft 76. Meisenheim am Glan: A. Hain, 1976.

Horapollo. *The Hieroglyphics of Horapollo*. Translated by George Boas. Bollingen Series 23. New York: Pantheon Books, 1950.

Timotheus of Gaza. *Timotheus of Gaza On Animals*. Translated by F. S. Bodenheimer and A. Rabinowitz. Leiden: E. J. Brill, n.d.

All quotes from the Bible are from the Revised Standard Version, Oxford: Oxford University Press, 1973.

1. *We begin first of all by speaking of the Lion* . . .

Aelian *H. A.* 5. 39, 9. 30; Plutarch *Quaest. conviv.* 4. 5 (670C); Horapollo 1. 19.

"intelligible tracks": Although this expression is somewhat awkward in English, I have chosen to translate the Latin *intelligibilis* throughout the text as "intelligible" since it is the equivalent of the Neoplatonic technical term νοητός, the noun form of which (τὸ νοητόν) means "that which can be known by transcending sensible things." Knowledge of this "intelligible" reality is close to the idea of Christian *gnosis* (see Clement of Alexandria *Stromata* 1. 93. 5).

"an angel with angels . . .": The view that the archons of the heavens were malevolent powers who would have resisted Christ's descent from the Father into Mary's womb had he not disguised his true nature by assuming their individual appearances ("an angel with angels, an archangel with archangels, a throne with thrones, a power with powers") was common among early-Christian Gnostic sects. The scriptural foundation for such a belief is found in Psalms 2, 24, 33, and 82; I Cor. 2:8, 8:5; Eph. I, 21. Irenaeus (*Adv. haer.* 1. 23. 3, 30. 12) mentions that among Gnostic circles the same story was told of Simon Magus's voyage to earth as well as that of the "Christ-power." For the secret passwords required of souls in their heavenly ascension through the inimical hierarchies, see the descriptions of Gnostic doctrines in Irenaeus (*Adv. haer.* 1. 21. 5) and Origen (*Contra Celsum* 6. 31). The closest parallel to our passage is found in the second-century tract *The Ascension of Isaiah* (ed. R. H. Charles), 10. 8–31. See also *The Epistle of the Apostles* 13; Justin *The Dialogue with Trypho* 36. 5–6; Tertullian *De carne Christi* chap. 14. Origen's view was that the progressive metamorphoses of Christ among the celestial hierarchies were accomplished for their salvation, but the idea was condemned by the Emperor Justinian (*Liber adversus Origenem* Mansi 9. 489–534) and the Synod of Constantinople (A.D. 543) as one of the errors of Origenism. This passage and similar ones in the chapters on the unicorn (XXXVI) and the adamant-stone (XLIX) are the only convincing traces of heterodoxy to be found in *Physiologus* (see Introduction).

"breathing upon the whelp": In other versions the sire's roar awakens the whelp. Abelard alludes to this tradition in one of his hymns:

> Ut leonis catulus
> Resurrexit Dominus,
> Quem rugitus patrius
> Die tertia
> Suscitat vivificus
> Teste physica.
>
> —*Analecta Hymnica* xlviii, p. 180, #170 (60)

11. *On the Antelope*

Pliny *N.H.* 10. 201; Aelian *H.A.* 7. 8; Oppian *Cyn.* 445f.

The scribes appear to have struggled with the name of this animal. The y-version has *de autolope*, the b-version *Autolops*. Among other Latin manuscripts the beast is found under the names *tholopha, talopus, calopus, utholphocha, authulphus,* and *autalops.* The Greek texts have περὶ ὕδρωπος, περὶ δρωπός, περὶ δρῶτος. Perry (p. 1092) believed that these names are corruptions of the Greek ὁ ὄρυξ, "a kind of gazelle or antelope in Egypt and Libya, so called from its pointed horns" (Liddell and Scott), about which Oppian (see above) writes. The legend was known among Arab writers, who called the beast *iachmur* or *jâmur* (see Sbordone, pp. 116–117).

The herecine shrubs belong to the heath family (Ericaceae) and include the heath, arbutus, azalea, and rhododendron, any of which might be meant by the *ricine* of the Latin text. The Greek word for the shrub is ἡ ἐρείκη and the Latin is *erice,* though the scribes both Latin and Greek were often baffled by this name, as the numerous variants in Carmody and Sbordone demonstrate.

"O citizen": The Latin is *polite nomine* a corruption of the tag πολιτευτὰ γενναιότατε (Pitra, p. 341) of the Greek text. In the following chapter on the Piroboli rocks the tag πολιτευτὰ (as also in the chapter on the beaver) is rendered into Latin as *epolitetota.*

III. *On Piroboli Rocks*

Pliny N.H. 36. 128.

The Greek πυροβόλος means "giving forth fire." The legend in Pliny, however, is told of the magnet.

"O citizen": The Latin is *epolitetota.* See the notes on the antelope.

IV. *On the Swordfish*

Aristotle H.A. 9. 37. 662b; Pliny N.H. 9. 47; Aelian H.A. 9. 34; Oppian Hal. 1. 338.

The ancients told this popular story about the nautilus, but it ultimately was attached to the swordfish (*serra,* ὁ πρίων). Legends concerning the dolphin's habit of accompanying ships at sea may also have played a part in the development of this story.

"Through . . . contrary": The text of y is deficient at this point, and I have therefore supplied the bracketed passage from b.

### v. On the Charadrius

Pliny *N.H.* 30. 94; Aelian *H.A.* 17. 13; Plutarch *Quaest. conviv.* 5. 7 (681C); Hermes *Kyr.* 3. 49; Heliodorus *Aeth.* 3. 8.

The Vulgate reads *charadrium* in the passage cited from Deuteronomy. The Greek χαραδριός refers to the thick-knee or Norfolk plover. The fact that this bird is not entirely white should not bother us since our author deliberately alters the zoological details of his stories to make them conform more evenly to the allegorical interpretation. The RSV reads "heron."

The ancient authorities limit the bird's sanative powers to cases of jaundice. Pliny states, "If one with jaundice looks at it, he is cured, we are told, of that complaint and the bird dies."

### vi. On the Pelican

Hermes *Kyr.* 3. 39; Horapollo 1. 11, 54.

The legend of the pelican, unknown to the ancients, is recounted of the ῥάμφος (the crooked beak) by Hermes and of the vulture by Horapollo. Moreover, there exists an alternate version of the pelican legend in which it is the male bird which spills its blood over the chicks. Since the allegory identifies the bird with God the Father and Christ, the act seems more appropriate to the male bird, but our version may simply be a case in which the original story has not been thoroughly altered to conform to the accompanying allegory. Although the pelican remained one of the most popular hieratic animals in Medieval art and literature, not all authorities accepted the legend. In his *Enarratio in Psalmum* 101, s. 1. 7–8 (Corpus Christianorum, vol. 40), Augustine comments: "Vos sic audite, ut si verum est, congruat; si

falsum est, non teneat. . . . Fortasse hoc verum, fortasse falsum sit; quaemadmodum illi congruat, qui nos
vivificat sanguine suo, videte."

VII. *On the Owl*

Hermes *Kyr.* 3. 8.
   The *nycticorax* of the Latin text (from the Greek ἡ νύξ,
night, and ὁ κόραξ, the raven) was clearly a species of
owl. Aristotle says of it, "The eared-owl is like an ordinary owl, only that it has feathers about its ears; by
some it is called the night-raven" (*H.A.* 8. 12. 597b).

VIII. *On the Eagle*

No ancient authority recounts the legend of the eagle,
which in *Physiologus* seems to be patterned after similar
stories concerning the phoenix, snake, and sun-lizard
and reinforced by Psalm 103:5. The eagle's dramatic
moulting must have supported the notion of its "rejuvenation." See Aelian *H.A.* 2. 26.
   The story of Susanna and the elders comprises chapter 13 of the Book of Daniel.
   See Abelard's hymn "Caelo celsius / volans aquila":

> Solis intuens
>    illic radios
> Summo iubare
>    beatissimos
> Visum reficit
>    pascit oculos.

> —*Analecta Hymnica* xlviii, p. 198, #202

IX. *On the Phoenix*

Herodotus *Hist.* 2. 73; Ovid. *Met.* 15. 392f.; Pliny *N.H.*
10. 4, 29. 29; Tacitus *Ann.* 6. 28; Aelian *H.A.* 6. 58; Dionysius *De avibus* 1. 32; Horapollo 1. 35, 2. 57.

"Adar . . . Phamenoth": There appears to be some
confusion in this passage. The Greek reads τῷ μηνὶ
τῷ νέῳ, τῷ Νησὰν ἢ τῷ Ἀδάρ, τουτέστι τῷ Φαμενὼθ ἢ
τῷ Φαρμουθί ("in the new month of Nisan or Adar, that
is, Phamenoth or Pharmouthi"). Since the allegory identifies the phoenix with the incarnate and resurrected
Christ, the author has attempted to fix the date of the
bird's rebirth during the Christian Easter season, that is,
during the Jewish Passover, which falls on the fourteenth day of Nisan, the first or "new" month of the Hebrew ecclesiastical year. Easter, the corresponding
Christian feast, is a movable feast and can occur in either
of the Coptic months of Phamenoth or Pharmouthi.
Adar, the last month of the Jewish calendar, appears to
have been added to account for the fact that Phamenoth
extends from the twenty-first of Adar to the twentieth
of Nisan. Latin copyists, who must have been quite un-
familiar with the Jewish and Coptic calendars, dropped
the mention of Nisan and added the gloss identifying
the Coptic months as "Greek."

The first Christian mention of this legend occurs in
Clement of Rome (fl. 96), *Epistula ad Corinthios* 1, c. 25,
and gained widespread popularity through Lactantius's
poem *De ave phoenice*. On the history of the legend, see
R. van den Broek, *The Myth of the Phoenix according to
Classical and Early Christian Traditions* (Leiden: E. J. Brill,
1972).

The notion that the bird is regenerated in three days
originated in *Physiologus* and is found only in it and versions dependent on it (Broek, pp. 214–216). Also, "In
the literary sources we find the phoenix as symbol of
Christ only in the *Physiologus* and in texts influenced by
it" (Broek, p. 160).

The mountains of Lebanon were famous in the ancient world for their aromatics, which may account for
the peculiar detail of the phoenix's visiting that country.
However, it should also be noted that this detail may
have arisen from the Greek homonymy λίβανος (in-
cense) and Λίβανος (Lebanon).

x. *On the Hoopoe*

Aristotle *H.A.* 9. 49B. 633a; Aelian *H.A.* 16. 5, 10. 16;
Horapollo 1. 55; Hermes *Kyr.* 3. 36. 25.

xi. *On the Wild Ass*

Pliny *N.H.* 8. 108; Oppian *Cyn.* 3. 197f.; Timothey of
Gaza, chap. 22.

xii. *On the Viper*

Herodotus *Hist.* 3. 109; Pliny *N.H.* 10. 169; Aelian *H.A.*
1. 24; Horapollo 2. 59, 60.

The female's hybrid shape, half woman, half crocodile, is an Egyptian element in the story and is not to be
found in the Latin and Greek analogues. Oral copulation is the consequence of the female's lack of genitalia.
In Pliny's version, the female bites off the male's head
inadvertently during the rapture of sexual pleasure
(*voluptatis dulcedine*), and the young, having been
hatched in the uterus, burst through their mother's side
in impatience at her habit of giving birth to only one
young per day.

Horapollo tells us that the Egyptians regarded the
viper as a symbol of a woman who hates her husband.
Similarly, in the bestiary tradition (see White, pp. 170–
173) this legend becomes an occasion for an extended
antifeminist harangue.

For an additional passage on the behavior of the
"viper" (more appropriately the snake), see chapter
XXXV, "On the Weasel."

xiii. *On the Serpent*

Aristotle *H.A.* 5. 17. 549b, 8. 17. 600b; Pliny *N.H.* 8. 41,
59; Aelian *H.A.* 9. 16, 66; Plutarch *On Isis and Osiris* 381B;
Horapollo 1. 2.

The self-regeneration of the snake can be traced to its
moulting habits, but the legend was probably reinforced

by the ambiguity of the Greek word for the snake's slough, τὸ γῆρας, since this word also means "old age" (see Lauchert, p. 16). Because of its powers of rejuvenation, the snake was sacred to Hippocrates and Hygeia, the goddess of healing. The sacred insignia of Asclepius, the Greek god of medicine, was a serpent. On the fourth nature of the serpent, see the notes to chapter XXXV, "On the Weasel."

xiv. *On the Ant*

Pliny *N.H.* 11. 109; Aelian *H.A.* 2. 25; Plutarch *De sollert. an.* 967F; Hermes *Kyr.* 2. 25.

"Flee, O man of God . . .": This catalog of heresiarchs is not found in the y-version. Sabellius (early third century) belonged to the so-called Modalist branch of the Monarchians, who attempted to preserve the unity ("monarchy") of God. Marcion (d. ca. 160) was a leader of a popular second-century movement which regarded the Christian gospel of love as utterly antithetical to the emphasis on the Law in the Old Testament. To Marcion, Christ as the God of Love had nothing in common with the Old Testament God of the Jews. Manichaeus (i.e., Manes, ca. 216–276) was the founder of a Gnostic sect originating in East Persia whose principal dogma was a dualistic concept of the primitive opposition of the forces of light and darkness. The heresy was rigorously combated, in particular by Saint Augustine. Novatianus (d. 257) was a Roman presbyter and rival Bishop of Rome opposed to the election of Cornelius as Pope (251) and the supposed "compromises" with paganism. Montanus (late second century) initiated the apocalyptic movement originating in Phrygia and spreading rapidly to North Africa, where it was championed ca. A.D. 206 by Tertullian for its rigorous asceticism; it came under attack by other orthodox writers and was condemned by the Asiatic Synods before A.D. 200. Valentinus (second century), an Egyptian theologian, founded one of the most popular branches of Gnosticism, based on an elaborate cosmology drawn from Ophite, Platonic, and Py-

thagorean doctrines; its principal teaching concerned
the redemption of mankind through a special form of
enlightenment ("gnosis") about the relationship of
God, man, and the cosmos. Basilides (second century),
an Alexandrian theologian of Gnostic persuasion, ap-
parently taught that all parts of the universe are good
and that all men will ultimately be rewarded with eternal destinies befitting their earthly lives. Macedonius
(d. ca. 362) was a bishop of Constantinople known from
the fourth century onward as a founder of the "Pneu-
matomachi," who denied the divinity of the Holy
Ghost. This doctrine was condemned at the Council of
Constantinople in A.D. 381, but Macedonius's actual association with the sect has been questioned. Donatianus
(i.e., Donatus) was the leader of a third-century schism
in North Africa whose theology denied the efficacy of
sacraments administered by unworthy ministers. Dona-
tism was refuted by Saint Augustine, among others.
Photinus (fourth century), a heretical bishop of Sirmi-
um, was a follower of Sabellius. According to Saint
Augustine, Photinus denied the preexistence, though
not the superhuman qualities, of Christ. His teachings
were condemned at the Council of Constantinople in
381. Arianism, one of the chief heresies of the fourth
century, rejected the divinity of Christ and was condemned at the Council of Nicaea in 325, but its followers
were protected by the Eastern Emperor Constantius and
Eusebius, Bishop of Nicomedia. Ultimately, after much
bitter controversy, Arianism was defeated in the East at
the Council of Constantinople in 381.

The absence of Nestorius's name from this list led
Lauchert (p. 89) to conclude that this version was composed before the Council of Ephesus in 431 when Nestorianism was condemned. But see the Introduction.

xv. *On the Siren and Ass-Centaur*

Homer *Odyssey* 5. 346, 12. 166f.; Ovid. *Met.* 5. 552f., 14.
88; Pliny *N.H.* 10. 136.

The reference to Isaiah is closer to the Septuagint than to the Vulgate.

Chaucer's Nun's Priest appears to have read this chapter from our author. He describes a happy moment of barnyard domesticity between Chauntecleer and Pertelote as follows:

> Faire in the sond, to bathe him merily,
> Lyth Pertelote, and alle hir sustres by,
> Agayn the sonne; and Chauntecleer so free
> Song merier than the mermayde in the see;
> For Phisiologus seith sikerly,
> How that they singen wel and merily.

—*The Nun's Priest's Tale* 4457–4462

xvi. *On the Hedgehog*

Pliny *N.H.* 8. 133; Aelian *H.A.* 3. 10; Plutarch *De sollert. an.* 971F; Timothey of Gaza, chap. 6.

xvii. *On the Ibis*

Aelian *H.A.* 10. 29; Dionysius *De avibus* 2. 8.

The title of this chapter is *De Hibice* (On the Chamois), but the remarks show that the author was thinking of the ibis. The title should be *De Ibide*.

The symbolism of Moses' outstretched arms was elaborated similarly in the early-Christian *Epistle of Barnabas* (ca. A.D. 70–100), chap. 12 (see Alexander Roberts and James Donaldson, eds., *The Ante-Nicene Fathers* [New York: Charles Scribner's Sons, 1899], pp. 144–145).

Throughout the chapter there is a play on the Latin *altitudo*, which can mean either height or depth (cf. "the high seas" in English).

The legend of Thecla is recounted in the *Apocryphal Acts of Paul*. "This book, Tertullian tells us, was composed shortly before his time in honor of Paul by a presbyter of Asia, who was convicted of the imposture and degraded from his office. The date of it may therefore be

about A.D. 160. The author was an orthodox Christian"
(M. R. James, *The Apocryphal New Testament* [London,
1820], p. 270). The episode involving Paul, Thecla, and
Thamyris and Thecla's miraculous preservation from
death can be found on pp. 272–281 (James). Inspired by
Paul's preaching, Thecla resolved to remain a virgin and
to reject her betrothed Thamyris, for which she was
condemned to death. Her steadfastness, however, preserved her from many horrid torments, among which
are the two mentioned in the ibis chapter of *Physiologus*.
Thecla's legend is mentioned again in chapter XXXI,
"On the Whale." The story of Susanna and the elders is
usually related in the Greek and Latin Scriptures as
chapter 13 of the Book of Daniel and was composed in
Hebrew toward the beginning of the first century before Christ. Judith's slaying of Holofernes is found in
the Old Testament Apocryphal Book of Judith, chapter
13.

XVIII. *On the Fox*

Aelian *H.A.* 6. 24; Oppian *Hal.* 2. 107; Timothey of Gaza,
chap. 6.
   "a rubbish pit": I have expanded this sentence in
accord with the Greek ζητεῖ ὅπου ἐστὶ τέμη τῆς γῆς,
ἐχουσα χοῦν, ἢ ποῦ ἐστιν ἄχυρα (Pitra, p. 351).
The fox looks for a rubbish or chaff heap (ὁ χοῦς), that
is, a hole in the ground (*scysurra*) where chaff and other
refuse are thrown knowing that scavenger birds will be
nearby. The text of b appears to be a rational attempt to
explain the cleft: "requirit locum ubi est rubra terra, et
uoluit se supra eam, ita ut quasi cruenta appareat tota"
(Carmody, p. 29). The Arabic translator explains the
cleft by saying that the fox seeks a foul-stinking place or
a putrid cadaver and lies down there (Lauchert, p. 87).
In one of the Syrian versions, the fox stretches himself
out in a place where there are beetles which draw scavenger birds to the supposed corpse (Lauchert, p. 85).

xix. *On the Peridexion Tree and the Doves*

Pliny *N.H.* 16. 64; Hermes *Kyr.* 3. 37.

τὸ περιδέξιον, Latin *ambidextrum*, is a name invented to correspond to the essential nature of the tree which drives off the dragon now from one side, now from the other. The story is told in Pliny of the ash-tree (bumelia), but he makes no mention of the doves. Hermes indicates that the tree grew in India (καὶ δένδρον ἐστὶν ἐν τῇ Ἰνδίᾳ λεγόμενον περιδέξιον).

"right . . . left": The play on the words right (*rectum*) and left (*sinister*) is more effective in Latin than in English.

xx. *On the Elephant*

Aristotle *H.A.* 2. 1. 498a, 9. 46. 630b; Pliny *N.H.* 8. 32; Aelian *H.A.* 8. 17; Strabo 16. 4. 10. 772; Diodorus Siculus 3.27; Plutarch *De sollert an.* 977D; Timothey of Gaza, chap. 25. 25a.

"tragelaphus": From ὁ τράγος, the goat, and ὁ ἔλαφος, the stag. "A kind of stag, with a beard like a goat, perh. the horse-stag" (Lewis and Short). See Pliny *N.H.* 8. 120. The author wishes to distinguish the *tragelaphus* of the Vulgate, Deut. 14:5 (translated as the "wild-goat" in the RSV), from the elephant.

The manner of capturing the elephant is told by Caesar (*Gallic Wars* 6. 27) of the elk (*alces*) and is first applied to the elephant by Diodorus (3. 27).

Wellmann (p. 41) rightly points out that the mandrake episode in the begetting of the young elephant derives from Jewish sources and would have been incomprehensible to the ancients, who had no concept of "original sin." The ability of the mandrake to aid in childbirth figured in Jewish popular legend. See Genesis 30:14–24, where Rachel's barren womb is made fertile by God after she receives Reuben's mandrakes. The Ethiopian translation explains why the elephants make use of the mandrake as an aphrodisiac: "He [the elephant] has no desire for copulation. And if he wishes to get a son, he goes at dawn to the vicinity of paradise etc." (Hommel, p. 88).

xxi. *On Amos the Prophet*

This chapter is supplied from Mann's edition of the b-version and is not found in the y-version, nor among the forty-eight chapters of the Ethiopian translation. It appears to have attached itself as a gloss to chapter XXII, "On the Roe," or perhaps to chapter XXIX, "On the Fig Tree," and then to have assumed an independent place in the text. In Mann's edition it appears as the third from the last chapter between the elephant and the adamant chapters.

xxii. *On the Roe*

Aelian *H.A.* 14. 16.
ἡ δορκάς, the roe or gazelle. The Latin text has both *dorchon* and *caprea* in reference to this animal.
"he is called God": This is a play on the Greek word for God (ὁ Θεός) and the verb "I behold" (θεωρέω).

xxiii. *On the Agate-stone*

Pliny *N.H.* 9. 111; Aelian *H.A.* 15.8; Arrian *Indica* 8. 11–12.

xxiv. *On the Oyster-stone and the Pearl*

Pliny *N.H.* 9. 107; Aelian *H.A.* 10. 13.
"Jacob blessing his son": These are actually the words of Isaac speaking to Jacob.
See the hymn "Ad laudem Virginis."

> Vellus rorem, virga florem
> Jam concepit, quem excepit
> Gedeonis conchula.

—*Analecta Hymnica* ix, p. 73, #90

and "Gabrielis vox de caelis,"

> Mater fui novo more
> Fecunda sacro rore
> .  Gabriele nuntio
> Genui sine dolore
> Reservatoque pudore
> Cum candoris lilio.

*—Analecta Hymnica* xxxi, p. 149, #152

Perhaps no other chapter better illustrates the tendency in the later versions (in this case Mann's b-version) of *Physiologus* to turn the originally laconic text into a verbose and meandering collection of sermons. See also Ammianus Marcellinus 23. 6. 85–87.

xxv. *On the Adamant-stone*

Pliny *N.H.* 37. 47.
Pliny points out a possible etymological origin for this stone's ability to repel all evil: "unde et nomen indomita vis Graeca interpretatione accepit."

xxvi. *On the Other Nature of the Wild Ass and the Monkey*

Plutarch *On Isis and Osiris* 376E–F; Horapollo 1. 16; Fabius Victorinus *Commentary on Cicero's Rhetoric* (ed. C. Halm), p. 223. 34–37.
Hommel (p. 39) claimed that the twenty-fifth day of the Coptic month of Phamenoth was regarded as an inauspicious day by the Egyptians, since on that day the wicked God Seth (who was represented as an ass) was castrated in his battle with Horus. Wellmann (p. 65f.) thinks that the story of the monkey is borrowed from a tale told traditionally of the cat.
"ysemaria": ἰσημερία, lasting an equal time, the equinox.
"The wild ass represents the devil": As this sentence is garbled in y, I have substituted the corresponding passage from b.

The bestiary (see White, p. 34) comments on the monkey's (i.e., the devil's) lacking a tail, "Although he has a head, he has no scripture," playing on the similarity of *cauda* (tail) and *codex* (book).

**XXVII.** *On the Indian-stone*

Pliny *N.H.* 37. 170, 190.
The original story was probably told of the water-stone, called ὕδρος, ὕδριος, ὑδρίτης or *enhydros* (as in Pliny). *Sindicus* means "Indian."

**XXVIII.** *On the Heron, that is, the Coot*

The *herodion* can be either a heron or a stork. The *fulica* or the *fulice* of y is the coot. Psalm 103 of the Vulgate has: "Herodii domus dux est eorum" and "Ciconiae domus sunt abietes." The RSV has: "The stork has her home in the fir trees."

**XXIX.** *On the Fig Tree*

Aristotle *H.A.* 5. 32. 557b; Theophrastus *Hist. Plant.* 2. 8. 1–3.
The story seems to have arisen out of a mystical interpretation of Amos 7:14. I have put "insects" where the author uses the more specific words *canopes* (κώνωψ), gnats or mosquitoes, and *culices*, gnats or midges.

**XXX.** *On the Panther*

Aristotle *H.A.* 9. 6. 612a; Pliny *N.H.* 8. 62, 21. 39; Aelian *H.A.* 5. 40; Plutarch *De sollert. an.* 976D; Horapollo 2. 90; Timothey of Gaza, chap. 14.
In place of "panther," the RSV has "young lion" in keeping with the Hebrew of Hos. 5:14. "Panther" is the reading of the Septuagint. Ancient readers would not have missed the play on the Greek word for panther

(ὁ πάνθηρ, i.e., τὸ πᾶν, all, and ὁ θήρ, a wild beast).
On the reversal of the panther's traditional character,
see the Introduction.

XXXI. *On the Whale, that is, the Aspidoceleon*

The germ of this legend appears to have originated in
India, where it is told in the *Zend-Avesta* (*Yasna* 10. 10–11)
not about a sea-beast but, rather, about a gigantic
horned monster over whom flowed yellow poison a
fathom deep. The story later appears in the supposed
letter of Alexander to Aristotle, where the setting is still
India, but the monster is a sea-beast. The Greek text of
this letter was originally composed in Egypt during the
third century A.D. The water-beast version of the story
is to be found originally in the Babylonian Talmud composed by Rabbi Rabbah bar bar Hana, A.D. 257–320. For
other accounts of the legend see C. Coulter, "The Great
Fish in Ancient Story," *Trans. Amer. Philol. Assoc.* 57
(1926): 32f.

The scribes had difficulty with the beast's name,
rendering it *aspidoceleon* or *aspidochelone* (Klingender,
*Animals in Art and Thought*, pl. 218a), *aspidodeleon*
(White, p. 174), and *aspischelone* (Carmody's b-version,
p. 44). The name seems to derive from ἡ ἀσπίς, a shield
or an asp, and ἡ χελώνη, a tortoise or turtle. A. S. Cook
(*The Old English Elene*, pp. lxxiii–lxxxv) argues that the
beast was originally thought to possess qualities of a
serpent more than those of a fish and that the name
should be rendered "the asp-turtle." Bestiary illustrations, however, leave no doubt that later tradition regarded the animal as a sea-monster. He is called the
whale (*cetus*) in the y-version and in some Latin bestiaries (White, p. 197), possibly because the Greek term
and its variants signified nothing to a Latin scribe or
his audience. Lauchert (p. 20) suggests that the *aspidoceleon* was juxtaposed with the panther in order to create a
contrasting pair of animals endowed with fragrant breath,
one a symbol of redemption, the other of damnation.

On Thecla, Susanna, Esther, and Judith, see the notes
to chapter XVII, "On the Ibis." The Book of Tobit relates
that Sara, the daughter of Raguel, was married to seven
husbands, each of whom was killed by the demon As-
modaeus before the marriage could be consummated.
Tobias, son of Tobit, married Sara and cast out the
demon as instructed by Raphael, by burning the heart
and liver of a great fish of the Tigris River.

xxxii. *On the Partridge*

Hermes *Kyr.* 3. 38.
The legend of the partridge may have originated as an
interpretation of Jer. 17:11. See Saint Jerome's commentary on Jeremiah 17:11 (Migne, *PL* 24. 820–821) and Saint
Ambrose's *Hexaemeron* (Migne, *PL* 14. 261–262), both of
which follow the partridge legend as found in *Physiologus.*

xxxiii. *On the Vulture*

Pliny *N.H.* 10. 12, 36. 151; Aelian *H.A.* 1. 35; Solinus
*Collectanea* 37. 14; Horapollo 2. 49; Hermes *Kyr.* 3. 1. 91–
93.
*Eutocius* (εὐτόκιος) means "aiding in birth." The ancient authorities narrate a similar story about the eagle
and the eagle-stone (ὁ λίθος ἀετίτης). The stone's virtue
is variously described: protection against sorcery
(Aelian), an aid in anchoring the nest (Horapollo), protection of the nest against snakes, and as a help in laying
eggs (Pliny). Aelian also states that the stone aids pregnant women to avoid miscarriage.

xxxiv. *On the Ant-lion*

Herodotus *Hist.* 3. 102; Strabo 16. 4. 15. 774; Aelian *H.A.*
7. 47; Hermes *Kyr.* 2. 25.
Our version of this story is somewhat garbled. The
Ethiopian (Hommel, p. 67) and the Greek (Sbordone,
pp. 73–76) versions agree that the animal has the face of

a lion and the hind parts of an ant, but they make no attempt to trace the hybrid form of the beast to the anatomy of the parents. Since the father is a carnivore and the mother a vegetarian, the ant-lion perishes for lack of food.

The translators of the Septuagint, in rendering the Hebrew *layish* as "ant-lion," seem to have drawn on knowledge of an Indian legend (cf. Herodotus 3. 102) concerning a gold-mining animal called the "ant" (μύρμηξ) by Herodotus. Classical writers attest to the existence of four-footed animals larger than foxes but smaller than dogs, which burrow into the depths of the sands of the northern Indian deserts and heap up mounds of gold around their holes. They are very ferocious animals who will kill any man attempting to steal their hoards. Perhaps because they were sand-burrowing animals, they were called μύρμηκοι (normally meaning "ants" in Greek) by the Greek writers. Herodotus seems to think that they resemble ordinary ants, but they were later considered a kind of feline, as in a passage of Agatharcides (second-century B.C. Greek grammarian); [see Müller (*Geographi Graeci Minores*, vol. 1, p. 158, par. 69)], who speaks of the μύρμηξ in a passage on the lions of Arabia. Aelian (*H.A.* 7. 47) also mentions the μύρμηξ along with tigers and panthers.

In his *Moralia in Job*, however, Gregory the Great offered his own explanation of why the beast is called the "ant-lion." For Gregory the animal is clearly an insect, not a mammal:

For the ant-lion is a very little creature, a foe to ants, which hides itself under the dust and kills the ants laden with corn and devours them when killed. Now myrmicoleon in Latin is either the ants' lion or more precisely an ant and a lion, because with reference to flying animals or any other small animals it is an ant, but with respect to the ants themselves it is a lion.

This explanation is accepted by Hermes (*Kyr.* 2. 25) and was adopted by Isidore in his *Etymologiae* (12. 63) and subsequently by the bestiaries which used Isidore as a source. See G. C. Druce, "An Account of the

μυρμηκολέων or Ant-Lion," *The Antiquaries Journal* 3
(1923): 347–364.

xxxv. *On the Weasel*

Aristotle *De gen. an.* 2. 6. 756b15, 3. 6. 756b31; Ovid *Met.*
9. 323; Plutarch *On Isis and Osiris* 381; Horapollo 2. 36;
Timothey of Gaza, chap. 39.
Aristotle protests against the belief that the weasel
gives birth through the mouth. The version found in
*Physiologus* (oral-conception and aural-parturition) is at-
tested by other early-Christian writers such as Clement
of Rome (*Recognitions* 8. 25, Migne, *PL* 1. 1384A) and
Timothey of Gaza, who mentions both versions. But a
hymn attributed to Thomas à Becket runs:

> Gaude Virgo, mater Christi,
> Quae per aurem concepisti.

The passage on the viper which in the b-version is
appended to the Weasel chapter arises as a commentary
on Psalm 58:4–5. It was transferred from its original
position here to become the "fourth" nature of the serpent in chapter XIII of the y-version, even though the
serpent in that chapter is said to possess only three
natures.

xxxvi. *On the Unicorn*

Ctesias (Migne, *PG* 103. 226); Aelian *H.A.* 3. 41, 16. 20;
Timothey of Gaza, chaps. 25, 31; Hermes *Kyr.* 2. 34.
The legend of the unicorn arises out of reports concerning the rhinoceros. Ctesias, a late-fifth-century B.C.
physician at the court of Darius II, is the first to report
on unicorns, which, however, he calls wild asses (ὄνοι
ἄγριοι). He tells us among other things that there were
animals in India famed for their fleetness of foot and
single spiral horn, which was an antidote for poison
when fashioned into a drinking vessel. Aelian adds that
this beast, normally fierce toward its own species, be-

comes mild and gentle toward the female during mating
season. Lauchert (p. 24) believes that an embellishment
of this idea accounted for the unicorn's legendary and
fatal susceptibility to the charms of a virgin. These
stories were ultimately used to gloss Biblical passages
(Deut. 33:17; Ps. 21:23 and 91:11) which mention the
μονόκερως of the Septuagint and the *unicornis* of the
Vulgate. On the history of the legend, see O. Shepherd,
*The Lore of the Unicorn* (Boston: Houghton Mifflin, 1930).

For Christ's descent through the celestial hierarchies,
see the notes to chapter I on the lion.

**xxxvii.** *On the Beaver*

Herodotus *Hist.* 4. 109; Aesop *Fables* (ed. Halm) 189;
Aelian *H.A.* 6. 34; Pliny *N.H.* 8. 109; Juvenal *Sat.* 12. 36;
Cicero *Oratio pro M. Aemilio Scauro*, frag. 4. 6–7; Apulei-
us *Met.* 1. 9; Horapollo 2. 65; Timothey of Gaza, chap.
54; Hermes *Kyr.* 2. 19.
See the Introduction.

**xxxviii.** *On the Hyena or the Brute*

Aristotle *H.A.* 6. 32. 579b; *De gen. an.* 3. 6. 757a3; Pliny
*N.H.* 8. 105; Aelian *H.A.* 1. 25; Oppian *Cyn.* 3. 288–292;
Ovid *Met.* 15. 400; Horapollo 2. 69; Timothey of Gaza,
chap. 4; Hermes *Kyr.* 2. 40.
"arenotelicon": a hermaphrodite or an androgyne,
from ὁ ἄρρην, the male, and τὸ θῆλυ, the female. The story
of the hyena's bisexuality survived Aristotle's protestations that it was untrue.

**xxxix.** *On the Niluus*

Aristotle *H.A.* 9. 6. 612a; Strabo 17. 1. 39; Pliny *N.H.*
8. 35–36; Plutarch *De sollert. an.* 966D; Oppian *Cyn.* 3.
407–432; Aelian *H.A.* 3. 22, 10. 47.
This legend is generally told of the hydrus (as in the
b-version) which lives in the Nile. The beast's name is
hydrus or enhydros (ἔνυδρος, or ἐνυδρίες in Herodotus

2. 72). Emma Brunner Traut has shown that the legends
of the niluus and the echinemon were ancient Egyptian
myths representing the daiiy struggle of the forces of
light and darkness. The enhydrus/crocodile battle originated at Buto and was connected with the worship of
Uto, while the echinemon/dragon version derived from
a separate cult of Atum-Rê at Heliopolis. Both versions
were ultimately based on the traditional antipathies of
Egyptian animals (the former being associated with
water, the latter with land). The Christian author of
*Physiologus* has very little difficulty turning these ancient myths into Christian resurrection legends.

"When the mud has dried": In other versions (see
Timothey of Gaza, chap. 43), the mud is used not as a
hard protective cuirass against the crocodile's sharp
teeth but as a lubricant to facilitate entry into its entrails.

The hymn "Unicornis captivatur" reflects awareness
of the hydrus legend:

> Hydrus intrat crocodillum,
> Extis privat, necat illum,
> Vivus inde rediens

> —*Analecta Hymnica* xxi, p. 36, #39

In the Waldensian *Physiologus*, however, the
hydrus represents the devil, who gains access into
man's heart and thus destroys him. See "Der Waldensiche *Physiologus*," ed. Alfons Mayer, *Romanische Forschungen* 5 (1890); 416:

Per l'idria laqual auci enganivolment lo crocodril es
entendu lo diavel que decep enganivolment l'ome, mas
aço que nos non sian deceopu de li engan diabolic
façan que Dio habite a repause cun nos per bonas obras.

XL. *On the Echinemon*

See the notes to the niluus chapter.

Echinemon: Gr. ἰχνεύμων; Liddell and Scott define
this as "an Egyptian animal of the weasel-kind which
hunts out crocodile's eggs (asp's eggs), Herpestes ich-

neumon." That the Hydrus/Ichneumon chapters follow
the hyena in *Physiologus* may reflect the earlier arrange-
ment of one of its sources according to sexual practice,
since both these animals (the ichneumon and hyena)
were regarded as hermaphrodites (see Aelian *H.A.* 1.
25, 10. 47).

### XLI. *On the Little Crow*

Aristotle *H.A.* 1. 1. 488b, 9. 17. 613a14; Aelian *H.A.* 3. 9;
Hermes *Kyr.* 1. 2. 14–16; Horapollo 1. 8, 9.

Among the ancients both the crow and the turtle-
dove were honored for their monogamy and chastity.
Aelian mentions that celebrants at weddings used to
sing a song called "The Crow" (ἡ κορώνη) which
praised marital harmony and that, according to legend,
hearing a single crow's call at a wedding was considered
an evil omen.

### XLII. *On the Ostrich*

No such story appears to have been told about the
ostrich among the ancients. Lauchert (p. 38) suggests
that the legend in *Physiologus* arises from explanations
of certain Biblical passages (Job 39:13–18; Jer. 8:7). In
Jer. 8:7 the Septuagint reads ἡ ἀσίδα, from the Hebrew
חֲסִידָה, the Vulgate *milvus* (the kite), and the RSV
"the stork." The Latin *struthiocameleon* (from Greek
στρουθοκάμηλος, i.e., ὁ στρουθός, the sparrow, and
ὁ κάμηλος, the camel) means the ostrich.

"that star called Virgilia": The author uses Virgilia in
the singular, presumably meaning a single star. The
more usual form is the plural, Virgiliae, meaning the
constellation Pleiades. The quote from Job corresponds
to the Septuagint, not to the Vulgate.

### XLIII. *On the Turtle-dove*

Aristotle *H.A.* 9. 7. 613a; Pliny *N.H.* 10. 42; Ovid *Amores*
2. 6. 11–16; Aelian *H.A.* 3. 44; Horapollo 2. 32; Hermes
*Kyr.* 3. 43.

The y- and b-versions represent two distinct traditions regarding the turtle-dove. The one emphasizing the bird's solitariness may arise from the Song of Songs 2:12, while the other stressing the dove's monogamy was common among the ancients.

xliv. *On the Swallow*

Among the ancients the swallow was regarded as the traditional harbinger of spring, as in Aristophanes' play *The Knights* (419): "A swallow! Spring is here." Aelian *H.A.* 3. 25 specifies that the swallow neither conceives nor hatches more than five eggs. The author of *Physiologus* may have deliberately altered the number of chicks to make this harbinger of spring a symbol of Christ, the harbinger of salvation.

xlv. *On the Stag*

Pliny *N.H.* 8. 118; Oppian *Cyn.* 2. 233–252; Aelian *H.A.* 2. 9; Horapollo 2. 87.
   The stag's use of water to draw the dragon (i.e., the snake) from his lair appears to be an innovation on the part of the author of *Physiologus*, since the ancient authorities agree that the animal performs the task with its breath. Horapollo, however, says that the deer flees the viper.
   "And David said, The high mountains are for the stags": the Vulgate has *cervus* (stag), but the RSV has "wild-goats."

xlvi. *On the Frog*

cerseus: from the Greek χερσαῖος, from or of dry land.

xlvii. *On the Lizard, that is, the Salamander*

Aristotle *H.A.* 5. 19. 552b; Pliny *N.H.* 10. 188; Aelian *H.A.* 2. 31; Hermes *Kyr.* 2. 36; Horapollo 2. 62; Timothey of Gaza, chap. 53a.

xlviii. *On the Magnet*

Pliny *N.H.* 37. 61.

Pliny uses the properties of the magnet and the adamant-stone to illustrate the Greek notion of the sympathy/antipathy phenomenon of the natural world (see Wellmann, pp. 81–94). The magnet is opposed to the adamant-stone, which if placed near iron will prevent the iron from being attracted away from itself toward the magnet. The juxtaposition of the magnet and adamant-stone in *Physiologus* may be a reflection of the sympathy/antipathy character of one of the book's sources.

xlix. *On the Adamant-stone*

Pliny *N.H.* 37. 57.

Saint Jerome (*Comm. in Amos prophet.* 3. 7, Migne, *PL* 25. 1124D) tells this story, claiming to derive it from Xenocrates (see Wellmann, p. 85, n. 226).

"It is called adamant because it dominates": A play on the stone's Greek name, ὁ ἀδάμας, unconquerable.

That the stone can only be found in the dark is a detail borrowed from legends concerning the carbuncle (ὁ ἄνθραξ). See Epiphanios *De gemmis* (*Opera omnia* [ed. 1623], vol. 2, p. 227).

"baldachin": The Latin text has *baldui*, a corruption of *baldachinus* or *baldicum*, a precious silk fabric interwoven with gold and named after the Italian for Baghdad (Baldacco), the city of its manufacture.

"Christ concealed his descent": See the notes to chapter I on the lion.

l. *On the Doves*

Aristotle *H.A.* 5. 13. 544a; Aelian *H.A.* 4. 2.

"Your lips": The quote, from the Song of Songs 4:3 in the b-version, is from the Septuagint, Ὡς σπαρτίον τὸ κόκκινον χείλη σου ("Your lips are like a scarlet braid"), and retains the Greek σπαρτίον as *sparcium* in Carmody's edition. The same passage in the y-version, however, is from a pre-Vulgate Latin translation: "Vestis coccinea labia tua, et loquella tua speciosa."

LI. *On the Sun-lizard, that is, the Sun-eel*

Aristotle *H.A.* 8. 17; Aelian *H.A.* 5. 47; Hermes *Kyr.* 2. 14.

The legend of the sun-lizard's blindness and recovery of sight arises from the animal's habit of sloughing off its skin (see the notes to chapter XIII, "On the Serpent"). Aristotle mentions that the "old age" is sloughed off by the gecko, the lizard, and the serpent and adds (*H.A.* 8. 17. 600b):

When the serpent begins to slough, the skin peels off at first from the eyes so that anyone ignorant of the phenomenon would suppose the animal were going blind; after that it peels off the head, and so on, until the creature presents to view only a white surface all over. The sloughing goes on for a day and a night, beginning with the head and ending with the tail. During the sloughing of the skin an inner layer comes to the surface, for the creature emerges just as the embryo from its afterbirth.

"whose name is the East": The Vulgate has "the East," but the RSV has "the Branch" in the passage cited. Jeremiah 23:5 (see also 33:15) reads, "I will raise up for David a righteous shoot."

Epiphanios (*Adv. haer.* 2. 53. 462) compares the Gnostic sect of the Sampsonites who called themselves 'Ηλιακοί with the sun-lizard and tells the story as it is found in *Physiologus*.